WATCHING GREAT MEADOW

WATCHING GREAT MEADOW

A Place of Joy ❦ A Place of Woe

September 2000 – May 2014

GORDON A. RUSSELL

2016
BAUHAN PUBLISHING
PETERBOROUGH * NEW HAMPSHIRE

ISBN: 978-0-87233-210-2

Library of Congress Cataloging-in-Publication Data
Names: Russell, Gordon A.
Title: Watching Great Meadow : a place of joy, a place of woe, September 2000-May 2014 / Gordon A. Russell.
Description: Peterborough, NH : Bauhan Publishing, 2016. | Includes bibliographical references and index.
Identifiers: LCCN 2015043703 | ISBN 9780872332102 (trade paper : alk. paper)
Subjects: LCSH: Natural history--New Hampshire--New Boston Region.
| Marshes--New Hampshire--New Boston Region.
Classification: LCC QH105.N4 R87 2016 | DDC 508.742--dc23
LC record available at http://lccn.loc.gov/2015043703

Book design by Kirsty Anderson and Sarah Bauhan.
Cover design by Henry James.
Author photo courtesy of Barry Wicklow, PhD, Professor of Biology,
Saint Anselm College, Manchester, New Hampshire.
All other photographs by Gordon A. Russell.
Printed at Versa Press.

BAUHAN
PUBLISHING LLC
PO BOX 117 PETERBOROUGH NEW HAMPSHIRE 03458
WWW.BAUHANPUBLISHING.COM
603-567-4430

Manufactured in the United States of America.

To the most utilitarian of animals: the beaver

(without whom, we wouldn't have any of this)

CONTENTS

Porcupine raiding the author's apple tree

WE NEED ANOTHER AND A WISER and perhaps a more mystical concept of animals. Remote from universal nature, and living by complicated artifice, man in civilization surveys the creature through the glass of his knowledge and sees thereby a feather magnified and the whole image in distortion. We patronize them for their incompleteness, for their tragic fate of having taken form so far below ourselves. And therein we err, and greatly err. For the animal shall not be measured by man. In a world older and more complete than ours they move finished and complete, gifted with extensions of the senses we have lost or never attained, living by voices we shall never hear. They are not brethren, they are not underlings; they are other nations, caught with ourselves in the net of life and time, fellow prisoners of the splendour and travail of the earth.

—Henry Beston, *The Outermost House*

PREFACE

IT WAS A SUNNY SATURDAY IN MARCH when I first met the Meadow. Looking out on the still and snowy marsh from Gordon Russell's kitchen table, I connected immediately with this place. One moment the suspended soaring of a great blue heron over Great Meadow's surface, the next a flurry of snow clumps avalanching off bordering maple trees—this environment so represented "everything New Hampshire," but could really be anywhere.

When Gordon handed me the journal he'd so vigilantly compiled over fourteen years, I took the bundle of pages with a serious sense of responsibility. I've edited many a manuscript, but I sensed this would be no typical portfolio project—this was Gordon's story as much as the Meadow's.

I soon discovered its message would be about a whole lot more than a man and a marsh. Over the next months, as we organized sections and retooled passages, I came to know the special personalities that reside in or visit Great Meadow. With every page, I felt like a passerby in a small town, its stories told through the words of an extraordinarily knowledgeable and sensitive individual—a rock-star naturalist who happens to have a talent for turning a phrase.

Gordon's day-to-day recordings punctuate the adventures and interactions among Great Meadow's diverse creatures and environs; truly, Mother Nature's inhabitants will come to life as characters you grow to know and care for. Yet there is a thread that gradually ascends in volume and tone from entry to entry—a deepening connection between Gordon and the Meadow that made me pause . . . that compelled me to seriously reflect on the many interrelationships nature is tasked with balancing every day and every night, year after year. After a lifetime blithely taking advantage of all the beauty and wonder our incredible state has to offer, I started to get it.

Wrapping up *Watching Great Meadow* was bittersweet—I think we both could have found "one more thing" to fix, or another entry to add. For me, it was much like the morning I drove my oldest to college: I knew I couldn't delay the inevitable letting-go, and I perceived I would never again take everyday dramas for granted. My stewardship had rightly come to an end.

These days, I understand there is a very real cause-and-effect to the way we treat our natural resources. Walking my dog along the beach road, I notice the swans on the marsh and know their purpose is more than to pose for pretty silhouettes. I see waterways and inlets as some family's home, rather than potential skating surfaces come winter. New Hampshire's mosquitoes are always a nuisance, but now I check my frustration because I realize there is a bigger stage and they too have an important role to play in (as Gordon calls it) Mother Nature's theatre.

This journal offers a peek under the covers of a wild habitat that is becoming more and more defenseless with every change of season. I hope it opens your eyes to awareness and touches your soul, too.

Laurie Peyser
February 2016

INTRODUCTION

A Meadow Welcomes.

I am an octogenarian, so I've had time for a variety of vocations and avocations over the years. They have included husband, father, grandfather, teacher, mentor, conservationist, volunteer, hiker, lecturer, photographer, trail builder, fisherman, hunter, piano player, traveler, gardener, veteran (USAF) . . . and all through these eighty-odd years, naturalist and watcher.

Watching Great Meadow is about sharing what I have seen looking down from my home and out on the Great Meadow marshland in New Hampshire during the years 2000 through 2014.

Homestead

In 1974, my wife, Barbara, and I moved into a New England settler's house that had been pegged and mortised in 1775. Prior to our purchase, the old house had sat mostly unoccupied for close to ninety years. It goes without saying that the long-neglected dwelling was barely livable, but somehow we managed; we were in love. By 1983, our watchful restoration was given due recognition, and the house was listed in the National Register of Historic Places.

At times, however, circumstances overcome the best of long-term expectations. In our first days there, only four farmhouses were located up the road past our place. By 1999, the building boom of the '80s and '90s had added 242 new houses, and we had lost what we enjoy the most: privacy and tranquility.

In late winter of that year we made the hard decision to leave our beloved house, along with its fine gardens. One day, a fateful opportunity came along when a friend telephoned to reveal that he had a parcel of land to sell: Would I like to look it over? Oh, yes. My wife and I have been smiling ever since. It was a forty-acre parcel in a neighboring town, and it bordered a one hundred-acre wetland, long known as Great Meadow.

Meadow

The story of Great Meadow began around eighty thousand years ago when a massive glaciation, known as the Wisconsin Glacial Episode, covered the higher latitudes of the northern hemisphere. The thickness of the glacial ice sheet would have measured at least six thousand feet. One cannot imagine the weight that would have pressed down on the underlying ground, or even begin to think about the grinding and bulldozing power of such a massive sheet of ice slowly moving across the land.

Eventually, the long season of cold finally ended and the glacier began to melt. When glaciers melt over tens of thousands of years, colossal, megaton chunks of ice are left behind. Gravity pulls on those great cubes where they rest, forcing the wet, muddy earth underneath to squish outward. (Think of pushing a baseball into a bowl of oatmeal.) When after a hundred or maybe a thousand years the ice cubes melt, the cavities left in the ground are full of water. Great Meadow was thus born from a chunk of ice.

Of course, native peoples knew of the place. And the earliest settlers from Europe found that Great Meadow provided nutrients in the form of meadow hay that sustained their livestock until fields could be cleared and planted with grasses.

When torrential rains flooded Great Meadow in the mid-1990s, the habitat of meadow grasses, aquatic shrubs, and emergent plants, such as cattails, rushes, and sedges, was forced to adapt. Many, especially the grasses and the majority of the rushes and sedges, lost the battle; too much water.

When a habitat evolves or is created (as beavers can create), such places fill with detritus over time. Thus, Great Meadow changed from a grazing garden for deer and a nesting paradise for dozens of songbird species to a water-filled world now open to all animals truly aquatic. Ducks, muskrats, geese, otters, egrets, herons, cormorants, mink, grebes, and amphibians all quickly found the newly created marsh.

Native fishes were last to find the marsh. Whether they arrived by swimming upstream (unlikely, due to the beaver dam barrier), as eggs or fry entering from incoming feeder streams (possible), or stuck to the feet of birds (even more plausible) isn't important. The fact they are here adds yet one more level of prey-and-predator to the marsh's food chain.

Great Meadow has come to define what a sanctuary is. A mature, diversified forest encloses ninety-eight percent of its shoreline. A three-acre field lies behind thickets of aquatic shrubbery at the far shore. Its primary feeder brook flows, unseen for the first tenth of a mile, through a complex of tussock sedge and black alder. A secondary stream is totally hidden within the large cattail colony at the north margin, but its important flow of water makes that nesting area ideal for ducklings and goslings in training.

Watching Great Meadow

In 1999, I designed and positioned our new home on a high ridge overlooking Great Meadow. The only meadow-side human habitation is ours. Two gravel roads, out of sight from the Meadow, carry only light traffic.

That's when my watching began. In a busy life, the actual time I've spent observing has taken up only a tiny fraction of the hours in any given month. The moments I've witnessed and written about in this journal are only fleeting incidents in the life of the Meadow—but in my eyes, those moments have been well worth putting pencil to paper, snippets of nature that someday a reader of my stack of journals might find interesting.

I watch Great Meadow's birds and mammals, its reptiles and amphibians. I watch its prodigious hatchings of insects. I watch the sunrises and sunsets. I watch the wonders presented by its weather and its precipitation in forms unending. I watch its aquatic gardens, with their vast sprays of blossoms. My ears listen constantly, channeling into my mind the choruses and calls of the wetland's wild inhabitants in their urgings to attract and procreate.

I have been watching Great Meadow's stage for fifteen years. Some degree of emotion and sheer fascination has spilled over into my descriptions, but my eyes are connected to an experienced body, and over many years of being observant of (and sensitive to) the natural world, I am most appreciative of these moments with nature.

However, all in nature is not well. The gestalt of the Meadow is developing frayed edges. I sense them through my close relationship with the erratic, seasonal pulses of life, in, on, and over this place. Great Meadow—at times subtly, at times clearly—shows the sum of the biological abnormalities that are accumulating under a changing climate.

Earth's climatic ills weigh heavily here. Over the course of living with the marsh, beginning in the year 2000, I've watched a disturbing decline in the numbers of species, both resident and migrant, that utilize the Meadow's natural riches.

Therein is a key realization that hopefully will be recognized by readers: If a family new to the area were to visit the Meadow today for the first time, they would certainly be impressed; it *is* an eye-catching place. But they would have no inkling as to what they were *not* seeing. There would be no hint as to the extent or variety of Great Meadow's past abundance. They would only remember what they see today—and that solemn, sobering fact has far-reaching ramifications.

For now, I share with readers these many gifts of life and beauty given to me by Great Meadow.

2000

Ospreys whistle. Herons croak.

SEPTEMBER 2

Osprey! It is perched on one of several dead trees that stand thick along the shoreline of Great Meadow. In the late afternoon, the slanted rays of the setting sun reflect directly on the bird's long, white breast. Without that emphasis, the mottled browns of its feathers would have blended in with the forest background.

Seeing an osprey never feels commonplace. Ospreys are large hawks, with long wings spreading five feet, sometimes more. They eat fish, and it is that activity that makes them so eye-catching.

Something in the water, probably a fish swimming just below, catches the osprey's keen eyes. The bird falls away from the perch in attack, but at the last second the prey either reacts to the bird's shadow or the osprey changes its mind.

At dusk, the osprey surrenders to the approach of dark and flies off to roost.

SEPTEMBER 3

Late this morning, an osprey with wings at full extension, "wrists" always bent, wheeled past my office window on a stiff breeze. Yesterday's bird?

This evening as I stand at the railing of my deck, with a hazy dusk settling in for the night, I squint to make out a dark lump on a dead stub in mid-marsh that shouldn't be there. Through my binoculars, I define the mass as a great horned owl. It's a bit surprising to see this nightly predator here because we have not heard its distinctive hoots since we moved here last month.

The owl's pose reflects its legendary all-knowing reputation. It turns its great head one way, then back, then the opposite way, on occasion exercising its ability to complete a half circle. The bird's ear tufts are long. "Horned" is an apt description for this grand hunter.

Experiments have proven the ability of owls to zero in with unerring precision on the sounds made by their prey. I squeeze my lips together and draw in a squeak. Instantly the predator swivels its head so its ears can read the sound. I am glad I am not a mouse.

An observation: Aircraft are vectored occasionally over Great Meadow on their flight pattern into the nearby Manchester, New Hampshire, airport. I have never noticed any marsh wildlife react to even the lowest flying planes—never that is until this evening, when this great horned owl actually twists its head and follows the progress of the low-flying object.

When twilight finally shifts into near darkness, it becomes difficult to distinguish where perch stub stops and owl begins. Perhaps the owl has stayed to watch the events of dusk on this evening. Scores of ducks begin to arrive, passing in front, behind, and just over the owl's head, wings set to land on the Meadow to participate in whatever the night has to offer. As they filter past, I can just see the owl's head swivel forth and back.

Do great horned owls ever attack prey at rest on the water at night?

SEPTEMBER 4

High humidity. The early morning is murky and misty with geese flying over the marsh, most heard but unseen. My spirit meter tips upward with the action. At this time of year, the major player of the seasons, summer, seems to become entrenched in itself, lackadaisical in a way, vulnerable to the season-in-waiting: fall.

This afternoon, sunset time, summer was indeed ambushed, literally displaced. The world of the marsh, its water, its sky, became tinted by fall's sly appearance in shaded colors, subtle hues belonging to its own identity. A tracing of fire-orange, drawn in round-edged patterns of clouds, spoke unquestionably of a new season. Less distinct in pattern, yet just as marvelous in color, were the clouds' mirror images reflected upon the marsh.

In fall, there is a thickness to colors, a certain quality that comes with the lower angles of sunlight on the green marsh islands, the pastures of pondweed, the rippled peaks of wind-tossed water, oh, even the great blue heron standing afar on a mudflat with sunset-facing feathers stained orange. By the time of this fine day's sunset, summer has been given a recess.

SEPTEMBER 5

Tall pine trees, in startling silhouette against a faintly pink pre-dawn sky, speak this morning of the season of coming coolness that announced its approach yesterday.

A great blue heron hunts in black water so dimly lit it barely supports shadows. Yet the hunter, standing now chest-deep, slowly telescopes its supple neck forward, forward, ever closer to some spot only its eyes can see, focused upon something to eat.

Stab! A hapless fish, caught exactly between head and tail, struggles. A few vertebrae-breaking snaps of the heron's beak render the catch limp enough to be deftly flipped so the fish's head points directly into the bird's gullet. The heron's throat muscles ripple, forcing the still-wiggling fish downward toward its final moments.

The first rays of sun touch the heron as it stretches—left wing, right wing, neck. A good start to the day (for the heron, not the fish).

In the evening, a bald eagle! Exciting to view such a grand bird. From our home eighty-eight feet above the Meadow, we can follow the chapters of its life.

SEPTEMBER 6

This morning the eagle rests on a great white pine, a major feature in the open spaces of Great Meadow. Long dead, the tree is a solitary object of stark beauty, with its array of bare limbs of various lengths and shapes. Standing high, it invites all birds that visit Great Meadow to land. It is especially attractive to large birds of prey. I call it the Perch Tree.

In the cool morning air (40°F), a mist has formed over the marsh. It is eye-catching. The mist grows in density and height above the black, rippled water, rising to a point several hundred feet above the Meadow to where the rising sun's first rays tint the top of it. It will be at least fifteen more minutes before the wetland itself receives the warming rays; meanwhile, the contrast in colors is magnificent.

Later, when the bright sunrise illuminates the surface of the marsh, the mist diffuses the rays, turning the surface milky white. I think of winter, recalling January days when the wind pushes blizzards of ice crystals across the wetland's frozen surface.

My recollection clears and the moment at hand becomes a mixed bag of loveliness, enhanced by the realization that there are many days of fall yet to be enjoyed. No reason to hurry time.

Now the sun has moved a bit higher and the distant Perch Tree is again visible, although without the wonderful eagle. Even so, the stark tree itself is an object of art. At a certain time during each sunrise, back lighting shows scores of spider-spun orbed webs. While many of the webs are constructed under the various dead limbs, some spinners have suspended them between the branches. Those delicate works in silk float like perforated balloons, adding to the austere beauty of the Perch Tree.

An osprey appears from nowhere and floats to a landing on the top-most spire. A breast feather dislodges and coasts away along a path of lazy arcs, reflecting sunlight in differing intensities as it disappears into the white mist below.

Around noon the osprey leaves the branch and quickly gains altitude, turning in tight circles upon a rising thermal. When it is over open water, the bird executes a steep rise into a stall, where it flaps its long wings in an attitude more vertical than horizontal. It bends its wrists at sharp angles, thus cupping its wings—an action followed by all birds that practice in-flight hovering, such as belted kingfishers and kestrels.

Folding its wings, the osprey drops in a tight spiral towards some prey. Just before impact, wings cupped again, it thrusts its talons forward and slams into the water. The prey escapes.

A rapid gain in altitude—droplets spray in all directions as the bird shakes off a feather-load of water. Back up to a hunting altitude then, almost immediately, the bird heads into another downward spiral, hitting the water.

No quick push to become airborne this time. I watch intently as the bird struggles to keep its head above water. Wings flail and stroke mightily as they reach for lift, but the weight of a fish gripped in its talons and the weight of water on its feathers have become challenges.

As the osprey repeatedly slaps the water with its wings, it eyes a mud patch nearby. The bird manages to row onto that shallow spot where it can get its wings above water. When it finally pulls itself into the air, I see that the talons of its left foot hold a very large fish. Gaining altitude with such a load requires enormous effort. Again, once in flight the osprey's first order of business is to rid its feathers of water.

The osprey heads back to the perch on the old dead tree, but its normal two-foot landing must be modified considerably. Coming in high over a chosen branch, it settles slowly downward like a helicopter.

The fish is long dead but the osprey struggles to keep the slippery catch on the branch with one foot, while trying to secure itself with the other. Over and over the heavy fish slides downward, and the osprey pulls it back up onto the branch—always with a wing flap or two. (Looking through binoculars from about five hundred yards away, I determine the fish was either a white sucker or an eastern chain pickerel.)

SEPTEMBER 7

Quite cool this early morning. Dancing water vapors are in motion over Great Meadow; yesterday's mists have moved northward. With little urging from the wind, the cloud just sits in place. Over a span of fifteen to twenty minutes, clusters of vapors slowly separate, prodded perhaps by the fingers of an awakening breeze.

At a little before seven o'clock the Perch Tree is once again occupied by an osprey. Same bird as yesterday, I expect, with a satisfied stomach.

SEPTEMBER 22

A morning worthy of a new season! The sky, washed blue, presents itself without a hint of variety. The air sparkles, touched on occasion by a few maple seeds spinning, spinning erratically in their earth-bound journeys and joined now and then by falling leaves.

Decorous fall has begun. Radiant red maple trees, more luminescent than remembered, are establishing their bright places along the far, eastern shore of the marsh. The low slanted rays of sunrise ever so slowly absorb the last remnants of dawn and proceed to spread their glory on the awaiting leaves.

Last night a skein of Canada geese dropped onto the Meadow from their flight south. They were the first migrants to be welcomed to the wetland's bosom for a temporary nurturing. The moon was in its waning quarter and the night air was so clear that the diffused moonlight softened the entire meadowscape. If only I could view such a scene from on high, through the experienced minds of the geese. Very possibly several members of that skein visited Great Meadow on previous flights, just as their parents had, and theirs before that.

The geese vocalized continuously until they skidded onto the marsh.

Once down they stilled, and to this point in the morning, no voices have been heard.

Hunting season alters the behavior of geese. How different they are from the summer inhabitants who dominate the marsh in early morning hours with their constant chattering. Without a doubt, gun barrels must have carried lethal pellets skyward towards these geese somewhere along their flyway. Stealth must be learned. Conversation has no place during times of danger, and fall is a time of danger.

SEPTEMBER 23

From the size and frequency of the ripples cutting through the marsh shrubs, it's clear that something other than ducks is moving behind a clump of winterberry bushes this morning. A black head appears. Only one animal moves with such slipperiness; only one animal bends its arcing body so slickly through its water world: the river otter.

At first, two otters appear—due to their spaghetti-like movements—it takes a moment to count a third head. Like a shuttle weaving its way through soft threads, these shiny-furred mustelids dive and surface, twist, turn, even somersault over, under, and between the multiple branches of emergent shrubs in seemingly joyful movement.

Three more otters join the party: six slick players, just at the marsh's edge, pleasing my smiling eyes. A family perhaps, or as often happens with otters, a joining of families.

I found an active otter den downstream along a tributary brook three years ago while I was surveying the major brooks in the watershed, and wonder if it is still active. At any rate, the joy of knowing that Great Meadow provides habitat for such delightful species deepens my appreciation.

SEPTEMBER 26

This evening the dusk has a damp feeling. A pre-sunset murk is suspended in the air, darkness finding room for itself well before its time. Somewhere to the west, a far, high hill acts as a final curtain to the setting sun, but above the hill the cloud mass parts, letting the day's final sunlight shine through, a sunbeam that illuminates the dark mist settling across the marsh, exploding Great Meadow into brilliant yellows and oranges.

SEPTEMBER 27

This morning, condensation created a huge, dirigible-like cloud that hovers just above the marsh surface. In barely-dawn light, all I can see is white, like a giant pillow laid upon the water.

With better light comes a slight elevation of temperature, enough to give the cloud mass a push northward. In the same way snow piles up in front of a moving plow, so the mist clears from the marsh's southern end, only to accumulate to the north. Like the inexorable up-thrust of mountains, the cloud folds in upon itself from both ends, forcing the central portion upward. It reminds me of a giant ice cream cone.

Unlike yesterday's eye-grabbing sunset, this morning's odd cloud formation soars to where rays of sunrise paint horizontal lines of pink across its summits, heating its molecules. The slight increase in temperature creates turbulence in the cloud's crown, causing the mass of water droplets to rush upward. Like the children caught up in the mesmerizing music of the Pied Piper, the whole cloud of mist follows the heated thermal and dances away.

OCTOBER 5

"They" (the annual foliage experts) have speculated that because the growing season this year has been perfect for deciduous trees (cool temperatures, ample rain), certainly the fall colors will be grand. Well, I've never witnessed a fall when the color has been anything *but* spectacular. I admit the shades vary season to season, but that fall treat—whether subdued or glorious—is always simply wonderful, leaving no room for quibbling.

Still, with the colors of October come the guns of autumn. Down from my house and across one thousand feet of meadow with its countless jigsaw surface pathways, sporting paddlers move their canoes through the pre-dawn light. Hunters, weapons at the ready (on occasion with an eager retriever manning the bow, his tongue lolling in anticipation), move into the recesses of grass. They pull their triggers. The hilly backdrop surrounding the marsh amplifies each explosion, which echo until the forest absorbs most of the explosive energy. The rest finds my ears.

How much the ducks and geese are bothered by the presence of the hunters or their triggered thunder is hard to judge. When shots are fired during the very first days of hunting season, those ducks or geese that escape the fatal sting of hot pellets often climb skyward and circle back directly over where they'd just taken flight. By that time, the hunters have reloaded and fire a second volley bird-ward. Neither ducks nor geese seem to repeat this strange behavior for the rest of the season.

The commotion in the marsh spells the end of a peaceful summer on Great Meadow, especially for the double-crested cormorants who have been temporary visitors from the coast. They've moved on, leaving me much indebted for their entertainment.

On the last night of their stay, I stood out on the deck and listened to their mews as they sat together high in the Perch Tree. While I had heard similar conversing from those birds, I'd only caught snippets of sound before, and never at night. The exchanges last evening were continuous, soft, and melodious; something I read "melancholy" into.

Now, by dawn, they are gone. What route will they follow back to their native sea? Will they dally? Or will they answer the call of the waters of their birth? I miss them already.

Far across the Meadow I see waters rippling, looking like the disturbances made by the gregarious cormorants as they team-fished for snacks. My binoculars reveal otters; certainly an acceptable substitute.

OCTOBER 17

The air is soft this morning. A thin cloud of mist spreads over the forest treetops like a light meringue. The stunning red of the maples has mostly disappeared, their leaves having lost their hold on their aerial stations. Once released, they sail back to the soil that nurtured them. Now, yellow is the dominant hue, with orange adding to the visual pleasure.

A pied-billed grebe swims out from behind a stand of waterweeds. This small brown water bird contrasts completely with the color now showing around Great Meadow.

With the pied-billed grebe the word "streamlined" can be completely disregarded. The bird is small and chunky, its head blob-like, its bill blotchy-colored (whence came its name). Pied-billed grebes are secretive in their movements, so when they do appear, that's a good moment.

This grebe is swimming only feet away from a frolicking twist of otters playing tag. Yesterday, the family of six performed close enough for me to see just how proficient they really are at pursuing prey, especially crayfish and fall fishes.

This is also the season of verbalizing blue jays. How obvious they make themselves, flitting there and here, practicing their falsettos and mewings, dislodging acorns from high in the oaks and trying to race them to the ground.

The jays seem very selective in their choice of acorns. I wonder if they choose only those that have been bored into by some tasty larvae? If that is the case, the acorns they drop could hit a rock and crack, and then a jay could easily get to the juicy larvae before it has a chance to wiggle away. That also might explain why, over and over, a jay follows a nut from treetop to ground.

OCTOBER 18

Around two hundred Canada geese came in last evening. When they departed this morning, the action out my breakfast window took precedence even over my chicken hash.

OCTOBER 19

Something new almost every day: In contrast to yesterday morning's misty grays and black edges, today dawns clear, with Great Meadow's trees wearing uniforms of yellow, orange, and tan.

Grebes are diving for food alongside a trio of otters. While I watch the otters curl their shiny black bodies in loops during their pursuit of breakfast, a great blue heron wings its way across the marsh, landing on the very cluster of buttonbush beneath which the otters are hunting. Not a minute elapses before a fish, in a frantic attempt to elude an otter's teeth, finds itself facing even a greater threat. The heron pulls its neck trigger, impaling the most unfortunate fish.

That heron's behavior is very interesting. Not every great blue takes advantage of the opportunity presented by otters disturbing fish, so it seems that this bird's behavior may have been learned from circumstance—that is, being in the right place at the right time, and acting on that experience later. The otters pay the heron no attention.

River otters are most distracting animals. By noon, three, maybe four, are porpoising across the Meadow, submerging fifteen to twenty feet, surfacing again, diving again, the lead otter stopping, head raised, waiting for the stragglers to catch up, greeting the laggards, and then back to the race. Soon they are out of sight in their haste to get to someplace else.

OCTOBER 20

With the low morning sun slanting its rays northward, the time is right to grab my camera bag and climb into my canoe.

Backlit spider webs catch my eyes first. The art behind spider webbing is endlessly fascinating, as is the silk spun to make them. Spiders make many varieties of silk, each kind specific to a single purpose, and webs serve many purposes simultaneously. No man-made material can match spider silk, in either tensile strength or durability. My eye decides, the shutter snaps, and now patience must be exercised—my old camera keeps the image hidden away on film.

The floating-leaf crowd is intriguing at this time of year. Pond lily leaves are used like sketchpads by battalions of aquatic leaf nibblers. It is not possible

Spider web with water droplets

to find a leaf that has not been munched on. The grazers vary in style, some gnawing at the leaves' surfaces, others content to work between them. The mosaics left on those thousands of leafy canvases can spin the imagination out of control.

Thousands of floating oval pads come close to roofing over the entire surface of the marsh, yet occasionally spaces become vacant. Each opening offers an opportunity to peer into the watery body, often clear as glass at this time of year because the multitudes of plant and animal plankton have disappeared now that the temperature has dropped. Just under the surface, water boatmen and backswimmers, both fascinating water bugs that use oar-like legs to chase down prey, dodge through the lacy branches of fanworts and hornworts. Together, they create a lovely underwater pantomime.

Bladderworts often join the troupe. They warrant special mention, however, because, while beautiful and graceful, they are meat-eating aquatic

Great Meadow's aquatic labyrinth of plants

plants in disguise. During warm seasons these floating snares, seen as tiny bladders among feathery leaves, are delicate in form and totally fascinating, but they are really vacuum tubes that suck in tiny aquatic invertebrates. In midsummer their miniature yellow blossoms, held high on stems, entice insects who service the blossom's dual purpose of obtaining food and dropping fertile seeds down into the marsh muck to await new life.

OCTOBER 24

I hear their voices long before they appear. Dropping on set wings from on high, a couple hundred Canadas sail in over my canoe this morning, landing at the north end of Great Meadow. Quiet at first, then sensing no danger, they begin to toss conversation about from bird to bird. Lively discussion goes on unabated for close to thirty minutes, increasing in both involvement and volume.

At that end of the Meadow there spreads an open pasture. Green grasses grow there, favorites of geese. However, it is hunting season, so these geese are wary; the talking continues, but the pasture is not to be trusted.

Five minutes pass. Then, rising above the grasses, fifteen geese launch themselves away from the din and fly directly to the pasture lunch counter. Another contingent rises up and settles in the field. Even though the entire flock of geese eventually makes it, at no time do more than fifteen to twenty birds make a move. They came as a flock; how was it decided which birds would go and which birds would wait? Caution, experience, evolutionary counsel—all contribute to the well-being of the flock.

How far did this flock travel last night before settling here? On a diet of mostly cellulose, it is nearly impossible to store much fat. Grass goes in one end, slips speedily along the birds' digestive tracts, and leaves as droppings, contributing little nourishment to its hosts in its passing. Geese are copious poopers, so they must eat a lot to keep up with the demands of migration. What is left in the birds' fat banks to allow them to continue?

After the last goose leaves for the pasture, I settle my canoe into a clump of winterberry shrubs, hiding as best I can. A quarter of an hour passes as I occupy myself thinking of my good fortune at simply being where I am.

Splashing, a couple hundred feet to my right. Before long I see a moving head. Otter! I am not surprised to see a second, followed by two more. I am well acquainted with the Meadow's family of these souped-up animals. They congregate behind a thickening of tussock sedge barely fifteen feet away. With camera ready, I wait. All I can see is a tangle of twisting black shapes, accompanied by short bursts of grunting sounds. I shift slightly to try a better angle, hoping their business will detract from their usual alertness.

There, seen through an opening between cattail stalks, is an otter squatting on an old muskrat lodge, with a black stick in its mouth. The otter circles the top of the mound, rubbing it with its mouth glands. The animal then drops the stick and immediately defecates a juicy mass of scat. Almost before it slides off the mound, a second otter repeats the action, sans stick, leaving its own deposit in the same spot.

For roughly ten minutes, the four otters spend time diving and coming up with mouthfuls of black muck, adding that material to what is surely a purposeful sign post, a territorial boundary mark.

I'm discovered! The canoe and odd shape do not belong here! Two otters, raising their heads and torsos high to get a better look, emit staccato-like barks, more snorts really, not very different from the alerts that deer give when confronted by a questionable scent or image. The pair, still curious and bobbing, comes closer. At about ten feet, not trusting what they are seeing, they bark once more and dive. That signal sends the others skedaddling also.

This evening, the gifts of living at Great Meadow keep coming. I find myself sorely lacking a vocabulary or the metaphors to truly express what my eyes catch and what my brain thinks they see.

During a sunset of deep color, tinted in hues of red and burnished orange, the wetland's animal activity continues. A heron, a great blue, flaps from bush clump to bush clump, not a typical activity for a bird with dangly legs and six-foot wings. But the heron has learned from watching otters.

When a playpen of otters corners fish, frenzy sets in—and for good reason: Such is the panic that a fish occasionally launches itself out of the water, often landing on land. Make that mistake in the presence of a heron and life is quickly shortened. Now the heron waits, ready, and its stomach is rewarded. A fascinating exercise in partnering seen repeated.

Meanwhile, the rich, dusky shadows of evening, added to the reflections off the forested edges of fall, provide a glorious backdrop for a martini. After just the first sip, two grebes commandeer my attention as they perform their head-down plunges in search of supper. The sky is a fairytale blue-pink, with only the pink washing across the marsh's surface. As the sun lowers to the horizon, the pink fades. Now on the darkening stage, the crowns of two poplar trees across the marsh glow golden yellow, like flames on giant candles, refusing to let go of the moment.

OCTOBER 26

Thirty-one Canadas on the Meadow turn their bodies in unison into a light breeze, and launch themselves southward, flying over the broadest section of the wetland. Both geese and ducks often take a route south through a gap in the forest where a brook leaves the beaver flowage.

One goose breaks rank, banks sharply, and returns to skid on the water. Pair, family, and flock relationships are strong with Canadas, so when one goose breaks rank, you search for a reason.

My deep calling in observing Great Meadow's body of life is trying to interpret, in a novice's mind, what I see. It is frowned upon by students of the animal and plant sciences for the layman to attribute human intellect or behavioral characteristics to non-human forms of life. For me, as an animal who has evolved from all animal species before me, I do not subscribe to the dictates that reasoning, decision-making, feelings, awareness, or emotions are physical attributes found only in human beings.

Furthermore, if I can't compare the characteristics, idiosyncrasies, peculiarities, or mannerisms observed in animals by using words that I feel are descriptive, what am I left with to share the wonders of my world? Joy is a human emotion—but is "joy" not what I see with a rolling scrum of otters? Is "joy" not what I see when a murder of crows is playing tag in a high wind? The numbing answer from the purists would be, "You have no proof the otters or the crows 'feel' anything."

Yet again this morning, the great blue heron is following the otters around. Am I being rashly anthropomorphic when I believe it spied the otters from some distant perch and "decided" (based upon previous learning) to fly to where the otters were because there might be food available? I call that a pure display of intellect (cognition, if you will), to take advantage of a previously learned situation.

While I appreciate that parameters for intellectual prowess differ for each species, in no way am I convinced that comparisons have any place in measuring differences or similarities between species. Intellect and physical design evolve as certain atmospheric and earth-based realities present themselves over extended periods of time. The abilities to cope, to adapt, to reproduce, are prerequisites for success. With mankind's proclivity to dominate and/or degrade his surroundings, along with the species they contain, I seriously wonder just how much abuse Earth's life forms can absorb.

When intellect is judged by longevity, certainly species like spiders, fishes, amphibians, and reptiles—all successful since the mid-Paleozoic Era some 350 million years ago—cannot be summarily dismissed as "dumb" animals. The way I figure it, the greater their longevity, the higher in esteem those species should be held. One should not compare intellect by holding humans as "the" standard.

(The single goose that left its flock and returned to the Meadow is now headed along the same flyway. Perhaps it simply forgot something.)

OCTOBER 30

A weekend of high winds spent itself out by stripping leaves from millions of deciduous trees and scattering them across the landscape to be recycled. The blustery visitor wasn't satisfied with mere wind, as several snow squalls rode in on the storm's caboose.

Words are feeble when needed to describe the actions created by the vagaries of severe weather. Today the Meadow's surface seems to be wearing white frocks—when any wave struggles to gain a respectable crest, the wind whips it to foam. When squalls of snow arrive, their burden of snowflakes is also at the mercy of the wind. Gravity, the accustomed master of airborne particles, holds no sway. The flakes, forced along horizontal paths as they fly, ride the gale like high-velocity crystal bullets. All structures—trees, bushes, boulders—in the path of the crystals end up wearing white frocks.

During the blast of cold, with all its turbulence and icy precipitation, a lone herring gull tacks its way, forth and back, taking up the challenge of the gale. In doing so, its performance becomes the personification of grace in the face of elements. Is it enjoying itself?

On the day before the wild weather's visit, just as the sun was awakening the air, two hundred Canadas gathered in the open water of Great Meadow. Each was neck-straight alert. No preening. No feeding. The birds, remaining in one section of the wetland, did not commingle as single flocks often do. Instead, they gathered in tight groups, members swimming betwixt and between one another like Dodgems at a fair. Their receptors may have sensed the approach of negative weather. It may well be time to continue south.

OCTOBER 31

At dawn, the conversation among the birds is sporadic and muted. As light intensifies, the geese have important matters to discuss. Decisions become final. Groupings are set. The first group off the water is up and away.

From watching over time I believe I can discern, among several clusters of birds, the one cluster that is about to take flight. Those in a particular group align behind a chosen leader. The voices among that cluster grow louder, necks become stiffer, the message "fly!" is passed—and that group is off.

Over a period of roughly ten minutes, the birds take off in stages. Interesting to note: Group Number One heads south, but Group Number Two steers southwest while Group Number Three veers off to the east, steering a wide circle around the Meadow before finally elevating over trees and disappearing off to the west.

The last group of twenty-one birds heads down the valley of the brook but problems apparently arise. One bird breaks rank and heads back toward the Meadow; somehow his action is "felt" by the entire flock. What was a neat formation of birds flying on track quickly becomes a jumble. Some geese turn right, some left—but each bird in that group returns to the wetland.

Winterberries for birds

The goose that broke rank and landed seems to have become a marker for all the returning geese. Once clustered on the water, there is much paddling and errant swimming. All are in high voice. No more than five minutes tick away before necks again straighten, bodies once more align, and takeoff is repeated. This time, apparently, everything is in order.

Strong sunlight follows those grand birds. My heart swells a bit with a sense of reverence and pure awe for those lives. So many unknowns.

Following the leaf-stripping, tree-exercising gale, the weather has begun to offer much softer conditions for the entrance of November. The blustery darkness has been replaced by scattered spills of gray, green, and burnt orange across the marsh. With afternoon light coming in from the west at low angles, the red winterberry fruits catch fire, making each bush stand out like a cluster of city lights as seen from a distance. In the sun's watercolor wash of faint pink, the lily pad colonies seem to blush in their exposure to such light.

A gang of crows materializes from who knows where, settling along the edge of the Meadow. They are strangely silent. They plod along the Meadow's edge, presumably seeking juicy aquatic bugs.

NOVEMBER 1

Sixteen Canadas appear from the north end of the Meadow this morning, slowly paddling into view. Six more appear from the cove across the way and approach the new arrivals. All together, the twenty-two birds make their way to the shore, pecking at whatever. A single goose stands erect on a flat of mud, preening. Ten minutes pass.

The lone goose on the mudflat completes what feather oiling and adjustments were necessary and begins swimming directly toward the band of geese along the shore. As this dominant bird gets within thirty feet or so, all action of the shore-feeding geese ceases. They get into the water and swim directly to the single bird. When the distance closes to a few feet, a directive is issued: "Take off."

Fifteen geese follow their leader into flight. The six geese that had joined the flock make a different decision and swim back to the cove. Conscious decisions are made to go or to stay.

Loops of goldfinches have been prevalent over the last three days.

The heron following the otters now has competition, as two more great blues are vying for the panicky fish, proving that learning through opportunity takes place every day.

NOVEMBER 2

Brilliant fall morning. Just as dawn awakens, the otter circus continues, much to my delight. They begin worrying fishes at six o'clock.

Today, they are mixing fun and games. I'm not sure, but I believe I count six otters. At first, they work the open water of the marsh. That activity is followed by pursuit of prey, which brings them very close to the shore (where I refuse to blink for fear of missing some action).

Next, they zero in on one of the muskrat lodges, where one otter after another scrambles up the pyramid to deposit musk. They seem to vie for who can leave the highest scent marker.

As the otter milling continues, the ever-watchful herons take note of the action. Three come to join the ruckus. Again they reap benefits. Fish that had escaped the otters meet their due with the strikes of the herons' beaks. Precious minutes for this watcher.

NOVEMBER 15

First thing in the morning, Gentle Ben, our extraordinary cat of seventeen years who came to us at our previous home as a feral animal, remains attuned to the natural world.

We are together on the deck. As is often true, when Gentle Ben sits up with his senses focused it's a good idea to pay attention: Rewards for following his alerts have come in both deer and bear sightings. This time it's wild turkeys.

One, two, then two more of those large birds fly out of the woods and up into a tall oak tree growing next to my deck. All told, eleven turkeys come to the tree. Apparently they had been feeding along in the woods, and drew the cat's attention.

These birds have some surprising aerial acrobatics to show me. From their high station, all eleven twist their heads down and turn them sideways. Apparently, they've spotted something on the ground under their perch.

My eyes gape as one turkey after another drops off a branch and, with wings only slightly open, spirals in an almost vertical drop to the ground. Such a sight! Unbelievable to see such agility and control in such a large bird.

After only a few minutes on the ground, all eleven birds fly up one by one onto a large branch offered by a neighboring great white oak. The assemblage of turkeys, now all in a line on the limb, preen together before settling in for a midmorning snooze, each head neatly tucked away under a wing.

NOVEMBER 16

A first: a red-bellied woodpecker. It demands my attention for several minutes as it hunts in the deep recesses of the old oak tree's ancient bark. How thoroughly it examines every possible hiding crack—places where beetles hide, crevasses in which a juicy caterpillar may have squeezed, or better yet, a plump cluster of insect eggs. Like all woodpeckers, its eyes are sharp and its beak's aim unerring.

No day seems to pass lately without otters supplying entertainment, all in concert with their frequent "friends of opportunity," the great blue herons.

NOVEMBER 18

Such a scene can only occur in November: clouds are stacked in layers of dark gray across the eastern horizon; the entire shoreline and the line of forest at the far marsh are streaked in sunlight coming in low and brilliant from the west. With the water's surface colored the richest of charcoal, every emergent shrub, stem, and withered leaf glistens in a stereoscopic mixture of colors, with yellows taking the lead.

Oddly, I'm reminded of snapping turtles. When sunlight reflects from that reptile's eye, I don't have to stretch my imagination far to see a starry galaxy offering a similar kaleidoscope of colors. Strange comparison, but who am I to question my recollections?

Late this afternoon, the day's departure seems to offer a wide variety of acts at Great Meadow, with stage sets changing most every minute. At this time, the water surface is no longer receiving direct light. The heaving, rolling heads of cumulus clouds are highlighted with pinks and oranges, each formation mirrored in the marsh. Looking at two skies at one time fills me with sheer awe. I catch myself holding my breath without realizing it.

Adding to the wonder are six otters, bobbing and diving like synchronized swimmers, bodies moving into pink and orange water. The instant an otter slices into a color, the water trail turns black; as the surface reorganizes, the erased color returns. It's as if the otters are drawing magical graffiti on a liquid canvas.

NOVEMBER 20

Ice covers the south portion of the Meadow. Canadas still linger to the north, feeding on grass in the pasture. A lone male mallard, his green neck luminous, cruises around in the open water, joined by five common mergansers, all males, each handsome from beak to tail.

The common merganser's beak is reddish, and blends into a neck that appears black until sunlight reveals a deep, rich green. In flight his body feathers are brightest white, while his wings are black blurs.

Within this species of duck, females do not take second place for feathered beauty. Their beaks are tinged red like the males' but their heads are distinctive, crowned with rich rust feathers that sweep back into a crest—the royalty of this species cannot be questioned.

NOVEMBER 21

The ice-covered portion of the marsh received a coating of confectionery sugar during the night. Wind-scalloped patterns of all sizes greet my eyes. A mink investigates the close shoreline, leaving tracks as evidence.

Blue jays are especially active this morning. The vigor of cold injects enthusiasm into their wild muscles.

NOVEMBER 24

Coldest night to date.

Close to one hundred Canadas are floating in the one area of ice-free water. After these birds fly to the pasture to feed, it takes less than an hour for that open water area to form a skim of ice.

NOVEMBER 25

Clear, cold: 8°F.

Great Meadow's ice is expanding, and the resultant cracking creates a symphony of eerie sounds. I hear the first harp-like notes from inside my home. These must not be missed. Once outside on my deck, my feet aren't too happy, but my ears are: clear notes, some sharp, some sustained, some modulations of several chords, all in an octave below middle C, building, building in intensity under the dictum of cold.

When the ice music travels to and from points all around the marsh, no two sounds are alike; the Doppler effect comes into play. If you listen carefully, the "music" emanating from a growing crack seems to travel away from or toward you. If the crack is approaching, the pitch rises. If the reverse is true, the pitch lowers with distance.

Tonight the contrapuntal music sends its icy notes into my warm bed.

NOVEMBER 26

Music over. The night's cold resulted in an ice layer thick enough that the marsh extends an invitation for a stroll.

Each year's first ice walk around the Meadow is a treat, yet my first steps on its slick surface are accompanied by both caution and satisfaction. The caution/satisfaction balance comes after I've walked some way without the ice buckling; the treat comes when the Meadow seems to have given its permission to trespass upon its roof and I am able to walk with assurance.

NOVEMBER 29

Forty-two wild turkeys saunter their humorous ways along the marsh's shoreline early this morning. They ramble along crooked lines, dawdling with a leisurely air, constantly plucking seeds.

The weather has turned disturbingly warm and I see it eating the fresh ice on the Meadow.

NOVEMBER 30

Warm rain: more bad news for the ice.

DECEMBER 1

Town tax bill time, but Great Meadow is putting on a show this morning that mollifies the annual stress.

Otters once again are the players this morning, with six performers now increased to seven. (Given the gregariousness of otters, the new member of the troupe was probably just passing by.) The thin ice keeps breaking under their weight, but these torpedo hunters go into action when they reach a shrubby clump of hard-stemmed plants.

Almost every time an otter surfaces, it holds a fish in its jaws. The marauders hunt for roughly ten minutes, eating fish after fish, then take a recess for ice antics. The scrum involves piling on, rolling, tumbling, and skidding across the ice. When the revelers reach thin ice and fall through, they scramble back and begin the entanglements all over again on a firmer stage.

Recess ends and lunch resumes. How many fish do they consume while I'm engrossed in observing them? Five to ten apiece? Maybe more. What is the metabolism rate of otters? It must be staggering. These otters' usual sleek forms seem to be much rounder than only a quarter hour ago. They take a breather from fishing, and the playgroup slithers over the ice, heading for the shore and vanishing among thick grasses.

The otters return in the afternoon but now in a party of four. How can you not watch them? They are back catching fish at the same rate they were this morning, but taking breaks to chin- and chest-box in between eating, reminding me of the chest-to-chest congratulatory bashes exhibited by football players. I wonder . . . do otters have egos?

I squint through the gloom as darkness begins to close down my spectator's bench: the otters are still feasting. A flock of Canadas that had been foraging in the pasture flies by in a perfect V formation, lifting up and over the forest to the south.

DECEMBER 3

This morning a black clump on the ice revealed itself to be a pile of snoozing otters. Not one stirred until early afternoon, when they all seemed to wake up at once. Their first act is to wipe their chins forth and back vigorously on the ice (fish scales must be very sticky). Then their fishing activity resumes. The gang is back up to seven.

With my new 15x50 binoculars I can make out that the otters are dining on horned pout and red-bellied sunfish, among other species. One otter snags a large pickerel—almost fifteen inches long! I realize the marsh harbors thousands of small fishes, but why don't they scatter when the otter predation is so intense? The seven hunters take only seconds to surface with a meal, yet the supply of fish remains constant.

Another question: When an otter surfaces with a fish, the prey never shows any life, never wiggles or struggles. From the moment the otters slither onto the ice, they begin to tear out chunks of their catch and swallow them. Even the large pickerel offered no struggle. Why? Might a toxin be involved?

DECEMBER 11

Thermometer registers -10°F. The ice extends a solid invitation to dance my way around the marsh. "Like walking on a cloud," I recall an intern saying last winter. With transparent ice, this analogy holds nicely.

In skims of snow close to the shoreline, I record fisher tracks and those of two foxes, one dragging a useless leg. I find a newly constructed beaver lodge in the geographic center of the marsh. I know of three lodges on Great Meadow, each showing recent repair and fresh cuttings. I assume that the wetland's resources are supporting two families. One family lives at the north end, the other at the south. The third lodge begs more questions: Which family built the centrally located lodge? Might the lodge be shared? Do beavers ever build a spare lodge? I never underestimate the all-around capacity or ingenuity of beavers.

DECEMBER 12

Rainbow! It's one of those cold-front-approaching days when air from the Arctic becomes bullish. Today flurries of rain led, followed by dark, continent-shaped clouds that dropped their concealed loads of tree-bending, boisterous wind, stirring up the entire marsh complex.

A sword blade of sunlight slants in from the southwest and pierces dark rain clouds. The prisms of water droplets refract the light, shattering it into colors. The rainbow's base sits firmly upon the ice and its spectacular arched array seems to purposefully stand its ground against the sweeping sheets of wind-driven rain.

DECEMBER 13

Ice skates arrive from L. L. Bean. The rain-smoothed surface of the marsh has frozen perfectly. My feet have not been laced into skates since high school. Unsteady at first, my stiff legs balking at unaccustomed demands, I push off, lift one foot and push off again. An old rhythm awakens in this old man, although hesitatingly at first.

A bit more daring now, and the blades begin to zing as they bite into the ice. Away I go wearing a huge smile! "Fifteen to twenty minutes is enough!" announce my untested legs. With a mixed bag of weather promised for tomorrow, the skate lacings might remain untied for some time, so despite my legs' protests, I add five more minutes.

DECEMBER 15

Four inches of snow and a period of freezing rain last night spoiled skating, but today is brilliant and cloudless.

Following a variety of "must dos," I manage late in the day to get out on the Meadow and check on the otters. I come across three locations where they found small openings in the shore ice and returned with food. I followed one track into the woods for one hundred feet, to where an otter found a cavity under a boulder. Nothing indicated he had come out. Do otters sleep in such places occasionally?

Farther on, indicators in the snow show two otter skid-and-slide trails leading to the water under the ice. A further search reveals a surprise: Roughly three hundred feet off shore is a patch of leatherleaf. The name of this wetland-loving bush is appropriate because its distinctive, brown-speckled green leaves are indeed leathery to the touch. It is a delicate addition to Great Meadow, with its bell-shaped, white flowers growing in lines along its stems. At the base of that cluster, spots of blood and fish scales reveal that two otters had been there. It becomes clear that otters know their hunting grounds so well they are able to swim under ice to a small opening three hundred feet away. Might they use echolocation?

Winter's bluster

2001

Bullfrogs boom. Bitterns gulp.

JANUARY 31

For three weeks, the ground has been covered with nearly ten inches of soft snow. I finally got to a project I conceived last fall: providing perches for birds out in the Meadow that would be viewable from the house.

First task: cut three holes through the twelve-inch ice cover. Second, provide a stable upright upon which to attach the perches—but there is a problem. Because of the accumulated muck from centuries of decomposition, keeping any support upright will be a challenge. I buy heavy-duty four-inch plastic PVC pipe and cut it into ten-inch lengths.

I push each length through the hole in the ice and pound them down into the substrate to just where their tops will not be visible after the ice melts. I select and cut down two types of perches: One, a straight twenty-foot dead tree with numerous branches that hopefully, will attract owls, hawks, red-wings, kingfishers, etc. The other two choices are live red maple trees, both of which should attract tree swallows and cedar waxwings. The thickness of each tree fits snugly inside the plastic sleeves. Now to wait.

FEBRUARY 6

Big snow: twenty-plus inches. An old-fashioned nor'easter.

My eyes are met with terraces sculptured like long, sweeping dunes this morning. This morning's snowscape looks like deserts I've seen from airplanes.

FEBRUARY 18

A new sighting for me at Great Meadow: at first, I see the animal stepping gingerly along the stonewall at the marsh's edge. Its color jogs to mind the word "fisher." A closer look with binoculars identifies a gray fox!

Ah, here is an animal few people ever see. It usually sports coats of silvery gray peppered with dashes of black, yet *this* stunning animal is chocolate in color, with the black markings very subtle. A wily hunter, it eats whatever the season has to offer—berries, chipmunks and mice, grasshoppers and crickets—and it often climbs trees to snatch a tasty fledgling from a nest or an apple from a branch. If a nap is in order, a fat tree branch serves just fine.

After last week's warmth and rain, large patches of marsh glisten with reflected sky. February's longer light brings the promise of seasonal change.

MARCH 17

Red-winged blackbird singing! This bird has found a snow-bound, ice-covered world at its meadow birthplace, one of the risks inherent in returning early to establish a favorable nesting site. Aquatic insects, its usual diet, are hidden under ice. The alternatives must be a variety of seeds, as the blackbirds eat them regularly at my feeders.

MARCH 18

Two turkey vultures.

MARCH 20

Not only does the snow depth measure an average of forty inches in the forest, but the Lion of March roared in with an additional twelve inches today.

No respite for wild creatures. Tracks show white-tailed deer are wandering along rural roadways. Easier walking but their search for food there will be fruitless. Not a good sign. They seem to be holding up well, but females carrying young are approaching a critical time.

MARCH 21

This morning, as two coyotes sally forth and back, following their noses from one muskrat mound to the next, their grace, purpose, and sharp instincts reflect healthy adaptive abilities for meeting the demands of a trying winter.

Still, the mid-March woods remain heavily laden with five or more feet of snow. Ice is holding on too, stubborn and strong and two-feet thick. The spring births of wild things will be late this year.

There is one non-marsh species I am concerned about. Two Labrador retrievers belonging to some owner to the north have been preoccupied with digging into muskrat lodges. I've found five mounds destroyed. For the past two weeks, despite my attempts to chase the dogs away when I see them, they persisted. I tried putting up a scarecrow ("scaredog," I guess) contraption—but the dogs just pissed on the post to mark it as "theirs"!

Today I found that they had dug into the side of the beaver lodge at the north end, coming unsettlingly close to the beavers' living quarters. When I went out on the Meadow to check, I could hear nursing calls from within the lodge; beaver pups. No doubt the threat from the dogs may well put additional demands on that ice-bound family.

I toss food scraps off the side deck for the dozen crows who arrive every morning for breakfast. The tidbits I toss out at dusk are usually snatched up by both the red and gray foxes.

MARCH 29

Warm rains are eating both snow and ice. Despite winter's grip, both hairy and pileated woodpeckers are drumming. They choose dead limbs on hardwood trees to tap rapidly with their beaks, broadcasting signals that seem to establish territories.

CRITTERS AT THE FEEDERS:

3 Black-capped Chickadees
30-40 Black-eyed Juncos
30 or more Goldfinches
1 Gray Squirrel
1 Pair of Purple Finches
3 Red Squirrels
1 Pair of White-breasted Nuthatches

APRIL 9

Finally, areas of open water.

One double-crested cormorant. This ocean-schooled bird circles the lower marsh, returns, and lands on a stump. I am thinking about that early arrival when I am distracted as an osprey whistles. It banks by one of my lookout windows. With a long up-sweep, it lands on a tree branch projecting out over the ice. Seen through my binoculars, the greenish-yellow glint of the bird's eye gives "clarity" a new meaning. Likewise, the purpose of its gray talons, enormous for the size of its foot's palm, is perfectly clear.

The bird flicks its head from side to side. Some image catches its attention. Birds of prey possess enviable eyesight. The eyes of ospreys do not face directly forward, as ours do. An osprey's eyes are set somewhat on the side of its head, so focusing demands a different combination of skills. They appear to use a lot of head movements to target prey and take final action.

Hunting techniques differ with ospreys: When hunting over water, an osprey often comes to a flapping stall in midair as it scans for fish below. If it sees prey, it partially folds its wings and drops into a dive as it negotiates the angle of attack. At the last split second before hitting the water, the bird thrusts its talons forward. The water appears to explode from the impact as the osprey impales a fish.

As I'm watching this bird come up empty-taloned after three attempts, two bald eagles soar into the Meadow from treetops to the south. They sight a convoy of common mergansers. One eagle makes a low pass over the ducks. The ducks dive, dodging lethal talons during several ensuing feints. Both eagles sport feathers of mottled white on their heads and tails, marking them as adolescents. They have yet to learn that catching diving ducks is difficult.

One of the eagles scores a fish and carries it to the Perch Tree. The second eagle flies to the north end of the Meadow beyond my sight. No sooner does the seemingly out-of-luck osprey rise from the water with a fish dangling from its right talon but the second eagle reappears from out of nowhere in hot pursuit. Through a series of sharp maneuvers, the osprey barely avoids the agile eagle but manages to maintain a grip on the fish.

Such purposefully sneaky tactics are common with eagles. Stealing prey can be easier than hunting for it. If the osprey drops its fish, the eagle will capture it as it falls; a deft maneuver to say the least. In this case, the osprey does not give up its hold and quickly gains altitude in tight upward spirals, an evasive maneuver that keeps the eagle always beneath in pursuit.

Who wins the game? The birds rise to a height beyond where I can see.

One river otter is feeding at a tiny exit hole in the ice, far from shore. How those animals find those openings remains a nagging question.

APRIL 12

Northern goshawk on the Perch Tree. Eastern phoebe back yesterday. Temperatures are supposed to warm to the sixties by the end of the week.

APRIL 13

Tree swallows course over the Meadow where aquatic insects must be hatching due to the warm-up. Five talkative Canada geese wander around atop the remaining floating ice, each, it seems, trying to out-mutter any competitor.

APRIL 17

An immature bald eagle appears this morning, about a year older than those seen earlier. Apparently, the river otters left bits of food next to their old fishing hole. The eagle, beautifully mottled in patches of light and dark tan, yanks and pulls at the ice-bound scraps for several minutes.

APRIL 22

Really the first day in this long journey from winter to spring when warmth has spilled welcomed melting over the area. Yesterday, the marsh ice gave up its stubborn hold.

This morning, an American bittern's gulping calls find my appreciative ears. An osprey arrives, dives, and snares breakfast. Great watching.

Wood ducks, mallards, and Canada geese—all are answering Great Meadow's welcome, as well as the strong urges of hormones. Swimmers wearing smart whites and sporting beaks striped with red, pursue the ladies who themselves are bedecked in the sleek feathers worn only by female common mergansers. The marsh waters are rippled by paddling feet, the air above punctuated by the voices of eager male ducks, all dressed in feathers that glisten in indescribable shades.

APRIL 24

Spring's first great blue heron makes its entrance, flapping by my windows with the same solemnity with which it left last fall. A belted kingfisher also cackles past.

Life is exploding in front of me, life in the pursuit of life. This marsh is an all-encompassing womb. It contributes something of necessity to every one of the species evolving within it. Both flora and fauna draw life and sustenance from it. The water warms and provides a medium in which life can hatch or grow. The mixture of living forms provides nutrient links in a complex food chain—one life for another and that life for another—where eaters get eaten in due course unless they top the food chain.

Such is the grand scheme of life in this marsh—natural, with limited interference from man. The wonder of it all! And I yet know so little of it. Yes, spring arrived today.

Bullfrog tadpole 'like holding jelly'

As if someone took the cover off a seltzer bottle that had been shaken, once the cap of ice is removed from the marsh, long-silent throats awake with vibrating vocal cords, filling the night with maleness and invitation. Peepers achieve decibel levels seemingly impossible for their size. Throughout the night, the geese hold a symposium of verbal intercourse.

Like a conductor trying to establish some reasonable cadence, an American bittern sends thumping, measured gulps into the maze of voices. Thankfully my night's sleep is fitful, giving me ample opportunity to simply lie awake as nature's notes of hope fill the night.

APRIL 29

Common merganser action this evening: twenty to thirty birds of each sex, diving and fishing cooperatively.

Males approach females in water-splashing, no-holds-barred trial runs at sex. When a male attempts to mount a female, his beak snaps at her neck as he tries to hold his position. Both birds go into a frenzy of wing flapping that excites the egos of nearby males who join in the game. When mating is the goal of the moment, both sexes show ample aggression.

A lone female makes a decision to paddle-fly away from the attention.

A bullfrog's boom opens the night's concert.

Her action is noticed immediately by several males who leave the "handy" females and paddle-fly in pursuit. The chase goes out of view, off into the Meadow to the north. Was it successful? I have no way to know, but ten minutes later it seems that the goal of the moment has changed: the mergansers, both sexes, are back together fishing.

MAY 5

A clearing along the edge of the marsh has attracted a northern oriole, a bird I missed last year. His distinctive whistle opens the concert from the forest. A killdeer follows, his sharp, unmistakable call ringing true. Geese provide never-ending percussion, while an American bittern sets the pace with measures uncountable of his low, long-winded bass cadenzas.

How many spring peepers calling this morning? How many rasping voices of pickerel frogs? How many coughing gasps from green frogs, all in a practice session for this audience of one?

MAY 6

Dawn arrives, coloring the morning air pink and blue, with orange shading on some edges. I realize every cloud formation in each sunrise and sunset will never be repeated. I try to look carefully at them all.

The morning also arrives with a new set of melodies from birds whose heritage is bonded to wet places. I listen with appreciation for all.

MAY 7

A silhouette of a bird soars high over the marsh. "Turkey vulture," spoke my brain. There is, however, something about the image that won't let my eyes stray. This bird is larger than a vulture, and after a turn or two he shows no upward slant of his wings. His primary wingtips spread much wider and the wingspan seems to reach much further. When the bird tilts and drops into a long, gradual sideways slide out over the marsh, a friend who happens to be with me, an accomplished birder, says a special visitor is in front of us: a golden eagle! The eagle tilts left and, when it turns, the rich bronze feathers covering its head and neck confirm its true lineage.

Scores of ducks scatter in reaction to the predator who may savor one of them for lunch. They recognize the hunter immediately; had the large bird been a turkey vulture, they would have paid scant attention. Keen awareness is a necessity in wetland life.

That golden eagle is a first for me! The inspiring bird searches for food while curving in wide circles around the fringes of the marsh.

A fisherman pushes a rattly aluminum canoe into the far edge of Great Meadow. The golden eagle shies and flaps away northward, leaving both friend and watcher with eyes agog.

I see two pied-billed grebes, a yellow-rumped warbler, a Carolina wren, and a northern oriole. The sharp whistles of the oriole override all competition.

MAY 8

There's a problem with the Meadow: the water level has dropped. I walk to the site of the beaver dam and the cause is obvious. Recent heavy rains have gutted a four-foot section out of the dam, approximately three feet deep. Worried? I'm not. I have faith in the engineering prowess of beavers.

MAY 10

Faith is comforting. It comes with experience. After watching the water level begin to stabilize, I go back to the dam site. The breach has been re-

A painted turtle searching for food

paired. In the few hours since the break, those incredible animals stopped a major leak. Many sticks were wedged in, across, and over; chinks were filled with mud; a few small stones were secured in places. The water level problem has been solved.

MAY 11

Black-crested cormorant.

MAY 12

A pair of yellow-bellied sapsuckers has chosen a dead oak trunk to excavate for a nest site. It becomes obvious that the primary excavator is the male. Once he defines the general shape of the nest hole, the female sapsucker pecks out the finish work. On occasion, both birds spend time inspecting.

Three days of steady drilling have widened and deepened the cavity, and now the interior wood is not as dense. The pair now acts as partners, working in shifts. The female enters and shapes the actual nesting room. Each pecking time lasts roughly five minutes. She then leaves, and the male enters. Apparently his duty is to clean up. He enters, and reappears with his beak filled with chippings and sawdust that he shakes away before heading back in for more. Once the nest has met parental standards, I expect eggs will be its first tenants.

I spy the first snapping turtle of the year sunning on a muskrat lodge this afternoon.

Thus far, this has been a spring without rain. April showers usually guarantee I will be greeted by wake robins, painted trilliums, Indian moccasins (or pink lady slippers), the scent-filled trailing arbutus, and of course, the saucy, delicate bluets and my favorites: the fringed polygala, also called baby toes. So delightful.

MAY 14

The black-crested cormorants now number five. One seemingly very young bird sports a light brown beak and mostly brown feathers.

I checked up on the beaver dam again. The repair has now been "professionally" smoothed off with the application of sod and chunks of aquatic mish-mash. One does not ask "Why?" when looking at structures engineered by beavers. Today small saplings, two inches in diameter and at least ten feet long, have been put in place against the repaired section of the dam as braces. Remarkable!

Four great blue herons.

MAY 15

Lines of dark, gray clouds are strung from the horizon today like a great curtain, putting an end to many consecutive days of sunlight. Now the forest is in deep shade, resting amid unabated music from feathered choristers. In truth those sounds have purpose beyond music. Some are meant as stern warnings to stay out of a certain space, while others are invitations to come closer so an introduction can be made.

The sun made an appearance just before sinking behind Candlewood Hill to the west. A single shaft of light ricocheted off the marsh waters and spotlighted what seemed like every newly formed leaf on every shoreline tree. That brilliant sight made my eyes dance. In mid-marsh, each aquatic plant's new leaves sparkled like tinsel. It's nothing short of a dramatic exit by a sun that had no role to play during the dark day.

MAY 19

For all the promise of yesterday's rain-swollen clouds, no leaks. Today is far too nice.

A single red-wing continues to make regular visits to my decks where I feed the winter birds. He comes even with me standing next to the feeders. We match sharp whistles, one for one. I concentrate on getting the call at the right pitch and duration. The red-wing has no such difficulty; he manages to pick up seeds, crack them, swallow, and whistle all at the same time.

The sapsuckers are alternating chores as demanded by their successful brood. A rose-breasted grosbeak has shown up, his breast and melodious song matching the definition for enchanting. The dam fixers, now on a well-earned leisure-break, are consuming mouthfuls of their favorite pondweed.

MAY 22

When American bitterns are within earshot, their strange call cannot be ignored, a call that is challenging to describe or imitate. One notable field guide describes the bittern's call as "a hollow, croaking, or pumping sound, like, 'oonch-a-tsoonck.'" Another reference describes it as a "slow, deep, 'oong-ka-choonk'—like a mallet driving a stake into mud." *The American Bird Conservancy Field Guide* writes about the call as "pum-per-lunk." *The Peterson Field Guide to Birds* says, "The pumping, the 'song' of the bittern . . . is a slow, deep 'oong-ka-choonk, oong-ka-choonk.'"

Having never been able to make sense of phonetically represented bird songs, the voice of the American bittern here at Great Meadow gives me no idea how or why one would attempt to match letters to the bird's call. Try as I might to compare them, the phonetic clues seem far different from what I hear; I'll leave that to experts.

The bitterns (and there are two pairs here) announce their presence many times during the day. After listening to hundreds of calls, I've noticed a couple of patterns that are repeated:

- When the birds call, they speak about once every ten to twelve minutes.
- They issue a series of calls at one location, then move.
- The voicing has a pattern—the first notes come in rapid succession and eventually trail off.

They are high-pitched calls (several notes higher than the calls of the barred owl). After pumping out the introductory, guttural notes, it is as if the bird breathes in a tankful of air, filling its lungs to capacity. Following the "fill up" there is a brief pause before the 'ka-choonk' sounds begin. The 'ka-choonks' vary in number—I've counted as many as forty-nine consecutive calls! The average number falls somewhere in the mid-thirties.

Like a clock's spring running out of tension, bittern calls clearly run down towards the end of each series. Also, about three-quarters of the way through their efforts, you can usually hear a slight hesitation, as if a quick breath is necessary to complete the flourish. The calls are certainly distinctive, and when visitors happen to be at Great Meadow and a bittern calls, their eyes get wider, their mouths smile, and their faces express disbelief.

MAY 29

Windows reflecting unobstructed flight paths can mean trouble for birds. This morning, a bird flies directly into a marsh-facing window. The impact sends chills through me. I stop writing to check.

I take a deep breath upon seeing a male sapsucker lying wings akimbo on the roof of the greenhouse, raising and lowering his head in obvious effort. His nictitating membranes barely close, barely open, all indicating a serious problem. The impact drove his beak downward onto his breast. I cannot tell if the beak is fractured.

The bird does not lose consciousness. His head moves slightly forth and back. He occasionally blinks. Ten minutes pass.

Knowing that his mate is sitting on eggs only thirty feet away gives me pause, but such pauses have no response—although questions do come: If the male does not show up to take his turn on the eggs, how long will the female stay if she does not hear his signal? If she leaves the eggs, will she stay away for the full time allotted during previous exchanges? If she returns and gives the signal but her mate does not leave the nest, will she wait and never enter, or, after a time, go in? If the day brings cold temperatures, will the eggs survive?

The stricken sapsucker's head moves, but unnaturally. He slowly turns it 90°, first to the right, then 90° back to the left. His eyelid movement increases in speed and frequency. Another quite rapid turning of the head, as if testing for mobility. Two 180° twists follow. A few snaps of the beak send feathers, impaled on it in the collision, flying away.

Now the bird seems to be gathering himself, realigning his body. He brings his wings into proper position. He stands, appears alert, makes three hops to the edge of the roof, and flies directly to a tree trunk adjacent to the nest hole. There, he pauses for about one minute. He gives the usual call, the female immediately leaves the nest, and in goes the mate. Only two of us know there had been an accident. May all be well.

JUNE 2

Just after one o'clock this afternoon, I spy a mature bald eagle on the Perch Tree, sitting slumped during a welcome downpour.

JUNE 3

Foggy with a steady rain, and there in the mist sits the eagle.

JUNE 4

The eagle left late yesterday afternoon, but was back on the Perch Tree with prey early this morning.

After many days of wet and gloom, a fine sunrise reveals lots of activity at the marsh, and the marsh full of color and action is a magical place. A pair of wood ducks and a trio of hooded mergansers are vying for one of my nesting boxes. These wooden boxes follow a standard design set by fish and game experts. Years ago, wood ducks easily found nesting holes in old trees. Harvesting practices have cut those same old trees to make room for new growth, and without the trees wood ducks have become scarce. By putting artificial nesting boxes in and around wetlands, the birds have begun to make a comeback.

The boxes are roughly twenty-four inches high, one foot deep, and ten inches wide. They can be placed on poles out in wetlands or, like mine, attached high on trees. The hooded mergansers are especially interesting to watch, as this species uses artificial nests as a viable option, if available. A female flies to a box, clings to its opening, peers in a number of times, then leaves. Oftentimes, shortly after one female leaves, a second takes her place at the box and repeats the inspections. On occasion, males fly to the box. They never peek inside.

Interest in the boxes inspires a series of fly-bys. A pair of hooded mergansers flies far out over the marsh in a wide curve, circles, and returns at high speed directly over a nesting box. This activity may be repeated several times, and the speed at which these birds pass makes the antics very exciting.

By nine o'clock the marsh beckons and I answer. With camera and binoculars, I escape duties that should be attended to but won't be, at least for now. I wake up the old fiberglass canoe, now with me forty-five years and counting. When I turn it over, there, tucked into the stern, is a giant five-inch-long millipede. Now those are worth recording on film. I can't resist holding and watching such an animal, one with so many moving parts. When disturbed, its first defensive response is to coil its chitinous segments, along with its scores of paired legs, into a protective curl.

I've learned to be patient when holding millipedes. Slowly, they begin to uncoil and the magic as they glide off can please the dullest soul. If ever there was a perfect example for the word "synchronous," this species embodies it. On its hundreds of legs, the millipede moves its tubular body along in a wave-like action.

As I push the creaking canoe out from underneath thick bushes, I hear a complaint. I tap the keel near the bow and a head slips out from under the gunwale: a muskrat. Apparently, the encounter causes no great trauma, as the animal slips into the water, swims a short distance, and stops for a snack of pondweed. I'll bring its secret dining place, grass blanket and all, back shortly.

Good to be out on the Meadow. The water is still cool enough to be relatively free of plankton. Bladderwort plants float in its watery spaces, weightless, feathery, giving no hint of their predatory ways. Floating pondweed is well on its way to covering wide sections of the surface with its mosaic of oval-shaped leaves. Already, leaf miners have nibbled their artwork in every visible leaf.

Red-wings find my floating about threatening. This is the season of egg laying. Nests must be defended. Great energy is expended to attract attention away from them. Males flash wing epaulets as they make themselves obvious by flying out and back onto a perch some distance away from the nest site, all the while whistling sharp calls. Females also get into the act by tweeting high, shrill notes. Members of this species are masters at concealing their egg-holding cradles. I know by this pair's actions there is a nest in the vicinity, but it takes me over a half-hour of maneuvering my canoe to spot it. One egg, now preserved on film.

One of the muskrat mounds has a new covering of fresh mud shingles and several new reinforcing sticks. The Meadow's muskrats put up with a variety of careless neighbors, showing indulgence to each in turn. Canada geese like to stand on muskrat lodges, preen, drop feathers, and leave scat about. Otters use the mounds for their games. Old snapping turtles find the mounds convenient for sunning, so they claw and scrape and heave their heavy bodies atop them, often balancing so that all four feet are free. Today, stealth pays off as I come upon another muskrat home where a large common water snake is coiled on a veranda-like ledge. I take care not to meddle.

As I drift, making my way between islands of aquatic shrubs, I chance by the periscope of another snapping turtle—only eyes and nostrils showing above the water. Despite his bulk and four large, paddling legs, not the slightest ripple reveals his presence. Only the tip of his head shows.

Tree and barn swallows ply their usual insect-catching aerial swoops and tight turns in a great exhibition of soaring freedom.

Back on land doing a variety of chores, a host of distractions keep my attention divided. I discover a new eastern phoebe's nest under the deck by my office. The nest is spilling over with fledglings, like the crown of a large muffin.

Since the collision with the window, I've aimed many a worried glance at the sapsuckers' nest, but today relief comes when I catch sight of both parents. They visit the nest every couple of minutes with morsels of food.

The male, apparently recovered, is fulfilling his role. He exits the nest with a beak full of what appears to be dead wood. It isn't. He is removing excrement produced by the young birds in order to keep the nest clean. Each time he flies to the same tree trunk, flicks away the material he has gathered, and repeatedly whips his beak forth and back on a branch as if the load is most distasteful. It may well be.

Next, the female enters the cavity, stays but a few seconds, and then flies to the same tree trunk as the male. She too wipes her beak, but even with highly powered binoculars, I can't see what she is wiping off. This process is now in its fourth day. Her action must have a reason, but what?

JUNE 5

The sapsucker hatchlings are being well cared for and today I can see what is on their menu. The parents' beaks contain large-winged insects, some dragonflies I'm sure, but I suspect stoneflies, caddisflies, or even fishflies are also being served. Hard to believe the long wings of sapsuckers could be efficient for capturing fleet dragonflies. After each parent exits the nesting hole, that curious clean-the-bill wiping action continues to take place.

Just at dusk, the trilling chorus of the Meadow begins: tree toads high in trees along the shore. Last evening, I heard only the occasional call. Tonight is apparently swoon night, with many toads on the make. I wish I knew what factors are in play that were not stimulants last night. Full moon?

JUNE 10

If one could look into a chunk of some solid metal and see the molecular dance in action, it might be similar to the scene across the half-mile of mist-filtered sunlight above Great Meadow this morning. Focus is impossible until a hatch of semi-diaphanous gnats pinwheel their multitudes skyward through the mists. Their extraordinary numbers remind me again of the diverse and productive womb in this bounteous wetland. When I view the backlit swarm through the lenses of my binoculars, my mind wanders, exchanging the swarms of gnats with an image of galaxies swirling in our universe.

Tonight, the bullfrogs and the tree toads make efforts to out-proclaim any competition. There seem to be two compounds of bullfrogs here at Great Meadow: the southerners and the northerners. (No drawl is discerned from those to the south, no Yankee accent from those to the north.) I expect volume is a measure of standing as well as of territorial control. Both camps seem to be equally loud and probably equal in numbers. Oddly, but perhaps by design, there are a couple of hundred yards of neutral space between the two camps where no groaning takes place. The line of demarcation does not ever seem to be violated; the peace between the clans is thus kept.

By midnight, the toads have whistled through their allotment of trills. The bullfrogs win decisively as they never stop bellowing throughout the night. How do those marsh creatures on the day shift sleep through the stentorian thunder?

JUNE 11

A turtle morning. When you are closely attuned to marsh cycles, it is clear there is a time for this and a time for that. The time for turtles has begun.

Five medium-sized female painted turtles wander around my home grounds this morning, answering to the ever-building internal pressure of developing eggs. Watching the turtles seek satisfactory nesting sites is troubling.

The females spend their entire day dowsing for possible sites with their noses, digging repeatedly to satisfy their sense of what will serve, but without success. At the end of their day, they make the difficult journey back to the marsh. Will they try again tomorrow? Or, if they are unable to answer the internal urgency to deposit eggs, are those eggs reabsorbed into their bodies?

JUNE 13

This evening, the angle of the setting sun across the marsh illuminates surface disturbances that resemble the ripples created by soft rain. Instead of raindrops, however, the slight movements are made by underwater creatures rising through the surface to either gulp a measure of air or to pluck a speck of prey. Their numbers, of course, are uncountable, but when one realizes that each dimple is created by an individual life, again appreciation builds for nature's fecundity. The value of each life form, like raindrops to the productivity of Earth, is basic to the heartbeat of this marsh. No life is insignificant.

JUNE 19

The sapsucker hatchlings have grown and so have their vocal cords, allowing them to produce a continuous din as their hunger reigns. To meet their demands, the time interval between feedings has shortened to about once a minute!

Again the adults have changed the diet. Yes, insect wings protrude from their bills now and then, but more often the morsels hang limp—juicy, nutrient-filled caterpillars. Food abounds in this forest but at such narrow intervals of feeding, what must be the level of skill to find food that quickly, knowing each life has its own methods for hiding? This is day fifteen. Feeding is constant from earliest light to latest. Big job.

Both wood ducks and hooded mergansers are parading their broods. This morning, I watch two sets of ducklings, both woodies. One crew consists of nine, but then a tenth showed up, apparently a "problem child" who had ignored its mother. With ducklings, one call from mother results in her charges immediately switching their paddle motors into high gear and planing across the water to her side. I watch as the undisciplined one, again and again, veers off on its own into thick grass. Its curiosity is on fire.

At the same time, a second parent and a brace of youngsters come bobbing along through a patch of thick pondweed. The errant duckling from Crew Number One emerges out of the grass and swims immediately into the midst of somebody else's kids. Uh-oh, perhaps a problem? The ducklings pay no attention to the brazen kid, but their guardian does. All I see is a very purposeful bobbing of the mother's head. Somehow that signal is obeyed, and the misfit speeds back to where it belongs.

JUNE 25

Yesterday, during a pelting rain, I spied the hungry head and gaping beak of a nest-bound fledging sapsucker thrust far out of its nesting entrance hole, ready for an incoming snack. This morning, all has been quiet. No sounds coming from the cavity, and no begging head at the hole. Was only one bird successful? Did that fledgling leave? In the end, did the successful young bird so dominate his siblings that they starved, or was just one egg laid in the first place? I'm not sure about sapsuckers.

Blue jays have mustered their young out of their nests and onto branches. The youngsters flap their wings constantly, trying to get their parents to feed them. One adult jay fills his cheeks full of black oil sunflower seeds from my feeder. He flies up onto one of the dead branches on the old white oak and disgorges the seeds along the limb for two young jays. Do they swallow them whole, having yet to learn how to separate the seeds' meat? I expect their intestinal tracts are forgiving.

Some thief stole into my garden last night and devoured every single young beet plant. Not a deer. Not a woodchuck. A few impressions in the soil left a hint—one culprit on whom to put the blame: a porcupine.

JUNE 28

The wood duck's half-grown ducklings are getting bolder, and are paddle-dashing around the marsh. They seem to have learned that insects are tasty but it takes speed to catch them.

While I watch the tiny racers, the mother wood duck grabs a large bullfrog tadpole. It takes several slappings and bill-squishings to finally still the wiggling catch and position it for easy swallowing, an action that takes even more manipulating and a lot of neck stretching to finally send the morsel on its way. I am surprised as I thought wood ducks were, other than bug eaters, vegetarians. Not this one.

Later I see half-grown hooded merganser ducklings grapple with bullfrog tadpoles. They struggle but win, squeezing the fat tads down their gullets.

JUNE 29

For an hour or so just after sunrise, I again see hundreds and hundreds of spreading circles across the surface of the marsh, quite evocative of falling rain. These circles create a similar wonder, but with a distinct difference: They're larger, and stretch across the marsh for as far as my eye can discern. They are made by tadpoles—thousands of tadpoles rising up to grab a gulp of air. Great Meadow's watery world demands too much of my brain.

Like a giant kaleidoscope restricted to shades of green and blue, the colors across the marsh this morning are truly eye pleasing. The greens seem to dominate at first, but when distance is added, the blues seem to absorb them. I'm provided with more depth perception for the scene by large flats of pondweed that are nearer to my lookout deck. As cumulus clouds move rapidly over the marsh, the mixing of the greens and blues in the changing intensity of light is truly kaleidoscopic; a rare exercise in painting for one's eyes.

At the bird feeders, a male red-wing escorts one of his young along the railing. While the young bird flaps his wings for attention, father cracks oil seeds, extracts the nutmeats, and feeds his eager charge.

Common grackles, goldfinches, white-breasted nuthatches, blue jays, red-winged blackbirds, rose-breasted grosbeaks—all make the feeder a sparkling place. (I don't feed Gentle Ben, our cat, on the deck any longer because the common grackles scarf down all his food in no time. Ben, conversely, is no threat to birds.)

Today, as on several recent days when the temperature has run high, one of the great blue herons at the Meadow can be seen in mid- to late afternoon, perched upon one of the muskrat mounds. There it does an unusual thing: It stands erect and simply relaxes its wings, letting them droop. The action opens its wrists as the wings flop open, exposing the undersides of all the wing feathers. The exposure is directly into the sun, always on warm or hot afternoons. Might this be a way to expose irritating mites to the sun's radiation? It could simply be an act of air conditioning, whereby body feathers normally covered by wings have the opportunity to dry.

JULY 5

A heron is standing on a stump, one it seems to have chosen for its own. This morning's toilet activities—preening, combing, wing stretching, and leg stretching—are undertaken with the greatest deliberation.

The bullfrog chorus booms away, still going strong even as the dawn ages. How do they sustain the energy to bellow eight or more hours, night after night? During the day an isolated bullfrog may express itself, but come dusk the entire bullfrog bass section begins to grind out a thunderous rumbling. All starts slowly. One frog takes the lead, but before a stanza is completed many competitors drown out its final notes.

I've listened long to the calls of these grand frogs. The male bullfrog community (and there are separate communities) is reactionary; that is, any one individual's call stimulates challenge calls from *all* the males in their section of the marsh. All members in the tuba section blast forth in unison. The volume builds quickly, reaches its crescendo, holds for several seconds, then fades quickly to its final measures. Total quiet reigns. If you are listening, you sort of hold your breath, knowing that egos rule. Very soon, some self-flatterer begins to boast about his prowess, only to cause a cascade of answering booms. Needless to say, the volume and cadence of the energized chorus at sunset shows little oomph come dawn. How do they have any energy left for sex?

If a bullfrog wants space, why, in a one-hundred-acre marsh, do they cluster together in a crowd? I suppose if one must voice-spar to prove a point, it would be a waste of time to do so all by one's self. Get into someone else's face and get some satisfaction. That may well be the credo of bullfrogs.

JULY 16

Dead Canada geese.

A week ago, about dusk, there was a message on my telephone. Steve, owner of Great Meadow Farm in Massachusetts, called to report that farmhands had found "dozens" of geese dead in his meadow pasture. Next morning early, I reported the incident to the New Hampshire Fish and Game Department and went to check, joined by the Fish and Game biologist. We found nine carcasses, all but one now black from decomposition.

Various causes ran through our minds. Botulism from the farm pond or a fungus picked up in the pasture by the grazing birds? Something in the dung from horses? Eight of the bodies were alongside a chain link fence. All of the bodies were intact.

Wanton killing rose to the top the possibilities. Coyotes? Probably not. The scene did not fit the way they operate. Foxes? Canada geese are too large. More likely the killers were the same two dogs that had dug into the muskrat and beaver lodges on the marsh back in March. They are both known for their tendency to wander. The dogs may well have panicked the geese, forcing them to either fly or run against a tall wire fence.

Six of the dead birds were half-grown goslings. One adult was found out in the open field away from the others, with puncture wounds in its diaphragm. Seven of the birds wore bands, so we put those in a plastic bag for examination at the Fish and Game headquarters. Later I followed up with a call but they had no answers.

JULY 17

This morning, a note of relief: Five separate pairs of Canada geese parents are parading their goslings through the pondweeds just out front of my watching spot.

The hatch is both small and late this year, not counting the carnage in the field. The members of this morning's crews vary widely in size. One brood of seven is mostly full grown, while other families range from half grown to much younger. At the same time, thirty Canadas cruise around the Meadow in a pack, seemingly with nothing much to do but take on airs of self-importance.

JULY 20

The *a capella* bullfrog singers are approaching the finale of their seasonal performance. Only the occasional hanger-on goes solo. The others have left the stage and are now content to settle in for summer, when they will become belligerent and try to eat everything that dares move within range of their trigger tongues and gaping mouths.

Nights are punctuated by the plinkers and plunkers: the green frogs. It's unproven, of course, but it seems from careful listening that most of the "greens" are not based where the "bulls" are, and with good reason. On two occasions, I have witnessed bullfrogs with mouths stuffed with large, green frogs.

JULY 28

Five cormorants on Great Meadow today. These birds go to roost very early, around six o'clock in the evening. So far, their arrangement on the Perch Tree follows a certain pattern whenever they come to the marsh. After watching for several consecutive nights, I can see that each of the five cormorants has one preferred roosting branch. The scene repeats every evening, within a time span of ten to fifteen minutes—simply a wonderful scenario.

JULY 30

A bald eagle appeared on the Perch Tree in midafternoon. Soon, a strong thunderstorm began drumming, followed by heavy downpours that added their own music. After the storm, the dark curtain across the Meadow lifted to show that the eagle hadn't moved an inch.

Now that the storm has boomed on its way, the eagle spreads its massive wings and lets them rest akimbo along the branch. Some sight—reminiscent of the heron! While doing all that wing spreading, whether for drying or just to have a nice stretch, the bird continually turns its head, this way, that way. Watching through binoculars, I'm most impressed by the eagle's beak, so sharp, so yellow, so efficient.

For a while there is no movement from the eagle, nor are any cormorants around. Then in late afternoon the eagle drops from the branch, sailing on open wings for roughly five hundred feet. With rapid and powerful acceleration, the predator gets up to speed. It maintains a level course around the forested edge of the marsh, then suddenly sets its wings and slips into a downward slide. A correction at the last second before it drops its talons and quickly thrusts forward. *Zap!* A fish, plucked from afternoon leisure and now impaled, is suddenly airborne. With late lunch in hand, the hunter returns to the Perch Tree. Only a second or two passes before it puts its sharp beak to work.

It is difficult to imagine possessing a body like an eagle's, one with such precise control over senses and muscles. From its high perch, this eagle *saw* its prey *under* the surface of the water at a distance of many hundreds of yards. It needed to make a whole series of determinations—approach angle, speed, direction the fish is swimming—and figure out how to not cast a shadow. During the final split seconds of ambush, strategic angles and lethal talons must be in sync. *Score!* On display is a combination of incredible skills that will be executed thousands of times over this eagle's long life span. Birds of prey have to be planners and thinkers: Every act of predation presents a different set of circumstances.

The eagle takes time to eat the fish, tearing a piece, looking around, all in no rush. It doesn't leave until late afternoon. All the while its assertive presence intimidates the marsh birds, in particular the cormorants. None are seen for the rest of the day until only one returns to the perch, much later than usual.

AUGUST 3

Nine cormorants on the Perch Tree at sunrise. Normalcy has returned.

Tree swallows, several dozen, stir the atmosphere above the marsh. Their numbers will increase as the species moves south.

AUGUST 14

Several flocks of Canadas take flight just after sunrise. They are answering to something. An idea? A calling? What?

AUGUST 21

A long-eared owl gives a concert of barely audible "coos" just before dawn's light, its usual time for calling.

Cedar waxwings. For three days now their numbers have been increasing, much later in the month than usual. There are hundreds this afternoon, banking, swooping, deftly picking insects off the water's surface—exciting action for watchers. Their flitting quest for food is a joy to watch, but there is so much more. Before me are two species of performers: insects, which are invisible to my eye, and birds whose choreography is dictated by the insects. I realize what I am seeing here on this fine, late summer afternoon is seriously dramatic, an aerial play where death is known but unseen, where these delightful cedar waxwings, by their flying magic, keep the loss of insect life hidden from all viewers.

With the sun slipping farther and farther northward at setting time, the lower angle of its rays becomes increasingly noticeable late in the day.

AUGUST 24

The tree swallows are up early dancing, continuing their enthralling, jubilant, migratory merrymaking. A hundred, maybe two hundred, maybe more, spot a suitably isolated cluster of buttonbush. Some signal to land is understood by all. The swallows drop down in waves, quickly covering the limited number of branches available. Many of the incoming ones land on other swallows. Through binoculars, I see only a mass shaped like a bush. In early and late sunlight, the "bush" twinkles as the light plays on the vibrating wing feathers.

Such action is not happenstance, as the scene is repeated again and again. Communication skills are practiced and learned. Just as quickly, off fly the swallows. Following a half-minute fly-around, they are back again. When the bird-filled bush becomes overcrowded, some stimulus fuses the mass into a unit with one collective mind. Some "fly!" signal ejects every swallow into the air instantly, as if they had been zapped with an electric spark.

What follows is also of note: at times the mass of birds launches skyward, losing all connection with anything resembling a flock—they simply scatter. At other times, the all-at-one-time flushing shows a distinct purpose; staying together in a tight formation undoubtedly deters predators.

The swallows practice their fascinating behaviors again and again. Probably vital to safety is how the swallows' actions change as morning moves to evening. At sunrise and for about two hours afterward, the clustering-on-bushes activity is intense. It begins to taper off by midmorning and the business of chasing insects gets serious.

At dusk, the swallows feed again, but not in any observable pattern. Some are just dots high above the Meadow, snapping up high-flying bugs; others skim the marsh, picking insects off the water's surface, sometimes hitting the water with a small splash as they misjudge the snatch a bit.

The guidelines for an orderly and safe migratory journey are being assimilated. With programmed destinations of the Gulf Coast or even Central America, it is impossible to imagine the countless circumstances, many adverse, that migratory birds face.

Just before dark I see seven black ducks feeding together. Do I know them from last summer?

AUGUST 25

The swallows are gone!

Cormorants? Seen, but hard to know if the group is still intact. I saw six fishing this morning.

Families of Canadas are more and more frequently in flight.

SEPTEMBER 4

Time does slip away. One glance across the Meadow and summer's top weather story is obvious: *Dry*.

The beaver dam, like all such barriers, leaks in various spots, which is an ecological benefit. The leaks allow a flow of water downstream even during the driest of times. By leaking in several places, some near the top of the dam and others in places near the base, the temperature gradient of the water is neither too cold nor too warm for life downstream.

Man-made dams are not ecologically friendly. Their water release either spills over the crest or drains out through some conduit at the base of the structure. If drained from the base, the water is cooler than the water above it; if spilled from the crest, the water is warmer. In both cases, delicate organisms downstream have to endure more extreme temperature changes. Below many such dams, the aquatic biota is sparse for those very reasons. It seems beavers, like most wildlife, are better at accommodating others in their shared environment.

SEPTEMBER 9

The face of Great Meadow continues to dry. The water level is down about fifteen inches. Considering that the depth of the water in ten percent of the marsh is twelve inches or less, much of the surface substrate is exposed.

A bad situation? The answer depends. Much wildlife, from small birds to herons, from small mammals to mink and raccoon, finds a wide range of delectables on and in the exposed muddy, weedy marsh dining surfaces. On occasion, nature turns "normal" conditions upside down. Such accommodations are not always guaranteed; otherwise, no species would develop the ability to adapt and benefit from the occasional rare change in circumstances.

SEPTEMBER 11

The marsh's water level continues to drop. The herons (I count six today) ply the marsh with greater frequency during these drought conditions.

Springtail insects on leaves

Today, they stalk almost the entire marsh with their slender legs seldom showing wet above their knees. They are eating well; a plus for herons, not so much for the animals they are swallowing.

SEPTEMBER 15

The course of the feeder brook is now the only source of water flowing into the marsh. The fish-eating cormorants, unlike the herons, are now restricted to pursuing their prey within the confines of the brook bed.

Viewing the marsh under these conditions is odd, but also exciting; the flatness of the whole marsh has become a giant invitation. With herons wandering around, I notice that their knees become rulers for my eyes. Even when several birds stand well apart, the water level on their legs is identical. An odd way to gauge, but I'll wager I can guess the depth at any given spot within an inch.

SEPTEMBER 19

A storm front rumbled through just before dusk this evening. The rain was sparse at best, washing only the outermost leaves. A rainbow formed due east and spread its standard array of colors about 15° below the zenith. Moments later, a second arced prism formed higher in the sky, its colors as always arranged in the exact reverse of the first bow. In the slant of this

A Springtail Experiment

I decided to find the answers to my questions by measuring the springtails' activity.
As unscientific as my measurements may be, I had to calculate some generalities.

Over the period of one hour, I recorded the following:

I take ten photographs. Each shows a random accumulation of insects at various locations in the mass, measured by a square-inch mark at each spot.

A later count shows an average of seventy insects per square inch in the ten photographs.

Very rough figuring indicates that there had to be well over one million springtails in the cluster.

light before dusk, the duckweed catches the muted rays differently and its greens become greener.

The marsh water level is still dropping. I checked the beaver dam today and found that the flat-tailed engineers had plugged two of the larger leaks.

OCTOBER 10

A walk to get the mail turned into quite an experience. The tiny, very ancient insect species known as springtails has always interested me, and what is this on my driveway today? Springtails! From a distance they appeared like a blackish smudge; an oil spill. Thousands, millions probably: a mass of popping life moving across a flat space in my driveway. Like an army they march, holding a clearly discernable shape, moving in one direction across the uncluttered surface. The front line forms a perfectly shaped, convex arc. As I stand here, I can actually watch the mass move.

I've seen hundreds of springtails before. They are often called "snow fleas," an understandable name as the tiny insects do seem to jump like fleas. Springtails get their name due to a snapping device that folds forward under their abdomen from their posterior. When that appendage is suddenly released downward, it pops the insect into the air, somersaulting it backwards several times its one-tenth-inch length.

I've long wondered the obvious: Since the appendage flips the animal backwards, how do they ever move ahead? But they do. What is their rate of forward movement? And, why don't they fly? Springtails are really ancient insects, getting their start before wings evolved. What is important about this insect species is what an ancient and intriguing form of life they are. They are countless in the New Hampshire woods, yet their wonders remain mostly unrealized and their bodies unseen.

NOVEMBER 6

Midday. Immature bald eagle on the Perch Tree, surveying the Meadow. Eventually it steps off and spirals down, eagle fashion, onto a shallow mudflat. With purpose, it stalks belly deep into the water and plucks some stringy mass off the bottom. Whatever it is, the eagle carries the mass back onto a mudflat, pulls it apart piece by piece, and swallows.

Next, the eagle flies over the marsh from top to bottom—twice—each time passing low over herring gulls and hooded mergansers. Somehow both species read the predator's intent, and pay no attention. Watching this scenario, where no bird is alarmed, and comparing it to those times when every bird on the marsh panicked and took cover when an eagle appeared, hatches the question: How do birds sense the difference between safety and danger? I haven't discovered a single clue.

NOVEMBER 7

Doubtful the drought conditions are responsible, but upon comparing last year's numbers of migratory birds moving through Great Meadow, I find my count is low for both ducks and geese. One wonders.

NOVEMBER 8

I wrote too soon. Last night, roughly five hundred Canada geese plus thirty-some common mergansers arrived at Great Meadow and greeted my eyes this morning. Add those numbers to this fall's sparkling colors, and it is no wonder my steak and egg breakfast gets cold!

By nine o'clock, a north wind visits and adds another dimension to the busy view. Gusts hit and overturn hundreds of lily pads. Their undersides are coated with slime that seems to condense sunlight into laser-like beams. Hundreds of such beams create a mesmerizing show.

Now, add the white breasts of half a thousand geese to Great Meadow's stage, and I am watching a once-in-a-lifetime pageant. Throw in the passing of a great blue heron, a red-shouldered hawk, and an osprey, and the word "humbling" comes to mind.

Shuttling over the ground just in front of me are several black-eyed juncos.

NOVEMBER 10

Four common mergansers.

NOVEMBER 13

Fifteen common mergansers! The high energy of these grand diving birds never seems to abate. How far did they fly before landing here from their breeding grounds in the far north, perhaps even Hudson Bay? Guaranteed, Great Meadow's ample fare will send them on their way with energy replenished.

NOVEMBER 14

The marsh grew a skim of ice last night. A pied-billed grebe and two herons passed through today, and half a dozen herring gulls have been colorful aerial visitors for half a week.

NOVEMBER 15

Too warm. Overnight, a wand of warmth freed up those water molecules that had been stilled.

Two weeks ago, a pair of unusual visitors wandered up from the marsh. I first saw them standing in my driveway, a stately pair of domestic geese: she, pure white and beautiful, he, orangey-brown, erect and handsome. Domestic geese, from who knows where, are apt to be belligerent, taking no nonsense from strangers. This mysterious pair is exactly the opposite.

At night they return to the marsh; safer? Every day, though, they return for a visit. To entice them a bit closer, I try slices of bread. A couple of times each day, I sit with a bucket of water (so they could rinse the dry bread before swallowing it) and break the bread into small chunks, at first tossing them quite a distance away, but slowly letting them fall closer and closer. The geese are building trust, and it's a joy to watch their body angles and neck positions over several days of interaction. The gander was the first to accept bread from my hand. The hen followed the next day.

The pair has definite likes and dislikes. The hen relishes apples; the gander, not so much. Grapes are a favorite of both. They grab the grapes and swallow them instantly. Watching those round shapes descend their long tracheal tubes brings a smile to my face.

This morning, I discover that both birds have considered it their duty to pick every cherry tomato growing on all my carefully nurtured garden plants. If a tomato had the slightest hint of pink, it was plucked. Only those high on the vine survived.

NOVEMBER 18

Despite their thievery, the domesticated guests are really endearing diversions.

Today a visitor comes by with his large dog. Once out of the vehicle, the dog spies the geese and gives chase. Any concern I have over the welfare of the geese is quickly allayed when both birds spread their great wings and soar out over the marsh. Such a wondrous surprise! They fly in perfect tandem, arrowing their way high over the open expanse. At the far shore, they set their wings and slide to a perfect landing. Any domestic goose I've met can flap its wings and run while giving chase to something, but for this pair to lift their heavy bodies up and away—that is totally unexpected.

NOVEMBER 19

The pair is back this morning. Over time, I've coaxed them with various rewards up the long stairs to my outside deck. The gander made the move first, followed two days later by his hen. They negotiate the stairs one at a time, one leg at a time—no problem.

No problem for them, perhaps, but having geese on my deck now becomes *my* problem. If, after going to the effort of climbing the steps, they find no reward waiting, the gander displays his impatience. He seems to have figured that if he pounds his head on the window facing the deck hard enough, he'll get some attention. Well, that attention has amounted to my installing a gate at the foot of the stairs!

NOVEMBER 21

For two days the gander has been missing. My binoculars reveal only the white hen standing alone, not moving from her spot in the weeds on the far shore.

NOVEMBER 22

She's swimming mid-marsh this morning. I call. She answers with a squawk and flies to my driveway. She eats, but shows no real interest; her perfect blue eyes, so intense, so direct, so piercing, seem to want an answer. Of course, that's my human side showing.

NOVEMBER 23

Cold. Much of the marsh is covered in ice. This morning the lone goose was standing balanced on one leg on the far shore, head tucked under a wingfold on her back. I look again in late afternoon but cannot locate her, and I have not seen her again.

I deal with the coming and losing of those unexpected visitors with unwelcome emotions. The strangers from some keeper, some barnyard, squeezed my heart. They were clearly a dedicated couple. They had trusted my hand and given me the warmest joy. On the day they took wing to escape the dog, I gasped in admiration at the sight of the pair flying strongly, close together, out, over, and far across the expanse of the Meadow.

You want to save them. Winter is lingering just over the hill. Calls to authorities go unanswered. Their memory is bitter sweet.

NOVEMBER 25

I take a walk to the lower end of Great Meadow to check up on any beaver activity I might find.

A shortcut to reach the Meadow's end runs along an ancient stonewall, which now has the company of many trees. From the time you catch a glimpse of the wetland through the trees, until you stand viewing the wide expanse of water and forest and low mountains and high hills in the distance, something inside keeps signaling your body to stop.

The grand scene before you demands your attention. Two objects draw you in: a massive beaver lodge and a grand boulder at marsh edge. The great rock extends an obvious invitation: "Climb up and sit a bit."

Once there, you are spellbound again. The vastness of the flowage, the beaver lodge that holds the creators of this space, the general scene of forest edges and of cloud shadows sliding across the ice, all would please any rock-sitter.

DECEMBER 26

Skating! Such a day for gliding. What's that overhead? A great blue heron, heading north. This is not supposed to be.

A single Great Meadow Cattail

2002

Wood ducks squeal. Tiny frogs peep.

JANUARY 6

With significant snow forecast for tonight, the grand ice-skating I've been enjoying will be over. I missed but one day on the blades, and this morning's glide may be the end.

Exploring wetland habitats on ice can't be beat. Great Meadow's uniquely vast cattail colony is a wonder to investigate, with the plant's giant seed heads holding both beauty and interest.

Some of the heads are wrapped in spider silk. Find a misshapen one, and it may contain the larvae of the cattail moth. Come spring, after the larvae has matured, the hip caterpillar appears, sporting bright orange bands that cover the length of its body.

If you are there during a warm winter's day, you may witness nature's most lethargic explosion: the cattail seed head slowly bursting. Several hundred thousand tiny seeds comprise each head. Each seed has its own silken parachute. As the seed filaments begin to expand, pressure builds up, and a small bulge becomes visible in the seed head, usually at its base. Once displacement begins, all those dry, slippery filaments lose their grip and any breath of wind begins to scatter the species' future skyward. Destination: the planet.

JANUARY 13

One cannot see silence, but this morning's gentle snow certainly epitomizes the idea of silence in motion. The crystals are unaffected by any current in the air. Only gravity directs their tumble, as their weight gently moves air molecules aside.

By watching a single sprig of eastern hemlock catch those hexagons of crystallized water, the flat surfaces of its small needles cushion the delicate

Cattail seed head slow explosion

flakes enough so that few are shattered. These are remarkable moments to watch close up. How rapidly the flakes accumulate. Within five minutes, the green of the hemlock sprig is covered.

"Austere beauty" aptly describes the leafless forms of deciduous trees in winter. Their starkness is transformed, however, when a tree's mass becomes laden with a frosting of wet snow. When most natural objects become bloated, they lose their inherent beauty. But a snow-covered tree takes on beauty and grand symmetry.

As the welcomed flakes intensify, the scene from my house changes minute by minute. The forest outline fades quickly and, as the flake numbers grow, all distant objects became opaque. Without an object on which to focus my eyes, my mind shifts into neutral and becomes restless.

JANUARY 19

Several days of putting miles on my old snowshoes; wonderful.

Two encounters today, one surprising, the other pleasing. Again, I flush a great blue heron that had been at the Meadow's outlet. It has to be the same heron I'd seen just two weeks ago, which had flown only a short distance down the brook before landing. Again, my heart goes out to the bird.

At this time of cold and ice, it should be wading along some fishy inlet at Chesapeake Bay. It must have some incapacity. How it has managed to live this long is a wonder.

The pleasing incident happens as I walk across the far field that borders Great Meadow. The abandoned field is in the flux of a natural succession, grasses yielding to shrubs and saplings. There in very readable snow under a cedar tree are the unmistakable prints of a bobcat. My attention narrows quickly: four rounded toes (they have five, but the tiny fifth seldom shows) and one large, triangle-shaped pad. The claws are retractable and only show in tracks when the surface is slippery or when the animal is chasing prey. I backtrack a bit, and am rewarded: blood stains on the snow and the end of a tail—chipmunk!

JANUARY 21

. . . and chance and good fortune just joined hands. The "chance" is that I'm reading Lawrence Kilham's *The American Crow and the Common Raven*. The good fortune is the activity I just viewed from my lookout window: three crows, flying erratically over the marsh.

When crows are spotted doing dizzy acrobatics at this time of year, you can expect the cause to be sex hormones tuning up for action. It seems no dive can be too steep, no somersault too outrageous.

Wait! The performers, who usually fly several hundred feet off the ground, are doing their act less than twenty feet above the marsh snow. My eyes pop when suddenly all three birds veer sharply down, hitting the soft snow with a visible "puff" of white.

Odd, I expected a mating action to begin.

With two birds in the snow with wings akimbo, a strange tail, longer than a crow's would be, points straight up. No movement for several seconds. When I put my eyes to binoculars, I see that the bird underneath is a sharp-shinned hawk! Both crows flip themselves off the hawk. The hawk immediately rights itself and flies off, low, across the Meadow. Both crows caw and fly up into the trees.

Eyes often misinterpret. In this instance, to my great surprise, one of the "crows" turned out to be a hawk. Crows badger hawks and plague owls, obviously a spirited game, but any enjoyment is probably one-sided. I can only assume (in nature that is always ill-advised) that the hawk had purposefully joined in the sky dance, but there came a time when the crows pressed some advantage and drove the predator into the snow. Here, the "winners" in the wrestling match scored but quickly released their opponent. One might have expected some purposeful injury to be inflicted upon the hawk. That didn't happen.

FEBRUARY 25

Winter continues weak. The drought, announced as the worst in sixty years, continues. Only a smattering of snow covers the forest floor.

I put on skates again yesterday. Found a coyote's deer-kill on the ice just south of the cattail colony. Also checked out a muskrat lodge that appears to have been torn into by an otter.

FEBRUARY 27

Awoken by the finest of alarms!

When the season of cold comes, night after night we slip into bed and draw the covers up chin-tight. Many of us even disappear under those covers, impatient for our bed's warmth to soothe our shivers. In such repose, our minds shift towards sleep, we being unaware, hopefully, of the good work our internal organs are up to. When the alarm clock jangles us back to wakefulness, our minds take over, our dreams are left behind, and a new day is underway.

This morning's alarm, looked forward to every spring, came at twenty minutes past five: "Konk-le-ree, o-ka-lee"—the first red-winged blackbird of the year!

The large marsh I live next to follows a lifestyle somewhat analogous to our own. It is at rest on the outside while internally its innards are never still. However, the alarm call of the red-wing represents a beginning for the marsh as well. There may be no visible sign of stirring under its icy roof, but marsh watchers know of the infinite patience of the alarm sounder, and it will repeat its call again come tomorrow's sunrise, and the next, and the next, and if need be, again and again until the marsh wakes up.

To describe the awakening of a wetland from its winter slumber one needs an entire dictionary of nouns and descriptive adjectives. It takes multitudes of words. The repeated calls of the red-wing reach into the marsh, into lives that are still eggs, animals that wiggle, have fins, have claws, along with countless others that one day will sparkle in the air above the marsh with their wings. The calls reach seeds, dormant until enveloped by warmth, destined to join species that hold promises of brilliant blossoms to be touched by the sun, to paint Great Meadow's canvas.

"Konk-le-ree, o-ka-lee." The red-wing's alarm is a proclamation to its mother marsh.

MARCH 9

Fifty percent of the marsh's ice cover is out. Spring is awakening, so time for one last venture out on what ice remains.

I spot a small hole in the base of a cattail seed head. I have to investigate. Inside the silky seed mass is a small cavity. In it are stored five nuts of the American beech tree; a private pantry stocked for a winter snack by the white-footed mouse. Situations like that don't just happen without reason. A mouse chose that one four-foot cattail stalk from among scores of others as suitable for some need. Did the tiny creature have those five nuts before it dug that entrance hole? The nearest American beech tree is at least two hundred yards away, up a steep embankment from the wetland where the cattails grow; a long trip for a mouse. Beechnuts are only slightly smaller than a mouse's skull; hard to believe that it could carry two nuts at one time. How could discovery, planning, reasoning, and memory not be taking place in the mind of that tiny mouse?

MARCH 15

A bald eagle, a mature bird, alights on the Perch Tree late this morning and immediately does a lot of feather shaking and wing drying. I figure it must have just returned from an unsuccessful hunt for fish.

A pair of hooded mergansers is showing interest in one of the duck boxes. While the handsome male, sporting the sportiest of costumes, stays on the water, the female makes a pass by one of the nesting boxes, swinging just over it into the woods then banking sharply back and entering the box. After fewer than three seconds, out she comes, coasting to a stop close by the male waiting on the marsh. What transpires between them? There must be some exchange but it's too soon, I'm quite sure, to agree on housing arrangements. Who's the boss?

In the afternoon, the eagle flies from its perch, heading directly for my window but then banking low along the wetland's edge, giving me a superb view from my high deck.

Eagles are often referred to as "majestic." Flying directly towards me, this eagle broadcasts its unquestionable power, its mana, over me then turns and banks away sharply.

Some time later I see the eagle on an emergent stub, gripping some unknown prey. It rips the meal apart, seemingly in some haste, perhaps because a visitor is badgering him. Perched on a stick only a yard away, one crow caws repeatedly while the eagle is eating. We all sometimes get interrupted at mealtime.

Whatever the eagle's meal, there is apparently residue left on its beak. The eagle hops off its low perch into water only inches deep. There it sloshes his beak forth and back, plainly with some irritation. Then slogging through some muck, it reaches a stick that stands vertical. Holding its head sideways, its mouth wide open, it cleans the inner portions of its beak. This is followed by more water sloshing. After that vigorous maintenance, it flies to a perch close by and gives a prolonged, head-to-tail body shake. Meal over.

MARCH 18

Snow! Close to five inches this afternoon. The water in the marsh is still near 32°F, so the snow just turns to slush. Swimming through that medium is tedious for the ducks, but what fun to look at the maze of twists and turns that are left to show their exact routes.

It appears as though one of the male hoodies is attempting to make serious advances toward a chosen lady. The process, as always with this species, is one of putting up with being ignored. When her suitor makes an advance, she quickly turns away and sidles up to one of the other males in attendance. Suitor Number One takes great umbrage and instantly dashes forward with wings flailing, snow-slush flying. He attacks with neck stretched to its fullest extension, and, like some missile, gives the competition a solid body whack. Ego repaired, he returns to "his" lady.

Understand, a patient observer like myself can be kept very busy. At one point during this mating game, the most active female hoodie spots another female sitting quietly. She lowers her head, aims, puts herself into fast-forward, and literally slams into the presumed competition. It's hard to know but easy to believe that the bully female felt great satisfaction from her act; the recipient, probably not so much.

During this kerfuffle, a fine mink scurries along the shore, close to where the ducks are demonstrating their likes and dislikes. The mink pays no attention as he pours himself into, under, and between every possible nook and cranny where some hapless invertebrate might be snapped up.

MARCH 21

Spring! But wait, winter has just arrived. The calendar is wrong again. Ten inches of snow has fallen, along with a shivering thermometer.

I turn my head toward the window and see an eagle on the Perch Tree. This bird, an immature bald eagle, looks ill kept. The feathers on both its head and tail are scruffy and dirty. Perhaps efforts to capture some prey under the current conditions of snow and mud explain its disarray.

The heating power of spring's sunlight certainly has been proven today, as the marsh went from snow and slush in the morning to an expanse of sparkling, floating ice crystals by sunset.

MARCH 22

The lion winds of March are back. The marsh surface is covered with running whitecaps.

In the morning, a red-tailed hawk lowers the count of red squirrels, skillfully plucking one off the deck railing. The phrase "it never knew what hit it" applies.

THE CHILLINESS HAS BROUGHT THE BIRDS BACK TO MY FEEDERS:

Black-capped chickadees
10-12 blue jays
The usual entourage of crows
Red and white nuthatches
30 or so redpolls
9 red-wings

MARCH 29

Following the snow, the hooded mergansers (five males and four females) are back and their interest in the nesting boxes has returned.

For the past three mornings, it has been the females taking turns sitting on a chosen nesting box. One of the females will pop up off the water and fly directly to the top of a box. That move does not go unnoticed by the ladies still in the water below. Almost immediately, her "ownership" is challenged. One, sometimes two females, fly to the box and try to dislodge her by jousting and bill snapping. Sometimes the queen is dethroned, sometimes not.

However, when a female who had actually been on the nesting box returns to the water, she makes her superiority known by initiating a violent pushing and shoving of those "sissy" females who were reluctant to join in the box-top game. During all the female posturing, the oh-so-handsome males simply twist and turn themselves 'round and 'round in the water, maybe waiting for the queen-of-the-box competition to sort itself out.

Something else I've seen with hooded mergansers: When a male mounts a female, he holds her completely submerged beneath him for a period of at least ten seconds. Having seen this numerous times, I believe the action is not a case of bullying; it is obviously just the way the act is consummated. The act, in each case, broadcasts a series of wavelets, the result of the rapid pumping of the male bird.

MARCH 30

Robin.

MARCH 31

Tree swallows over the marsh. Two male wood ducks. The phoebe male has returned. Two American bitterns calling while I'm canoeing.

APRIL 1

Mature bald eagle on the Perch Tree, sitting hunched in pelting rain early this morning. First spring peeper heard, late afternoon.

APRIL 2

Serious prenuptial negotiations seem to have finally begun for the hooded mergansers. This morning, eleven birds gathered in a small area of the marsh that is free of aquatic plants, and a lot of chasing took place, with the females doing most of the antagonizing.

I have no idea what they sorted out, but by late afternoon, the action changes and the males get involved. With a flurry of pomp, both sexes thrust out their chests to the max while bobbing their heads in the sharpest of motions—but it is the males who take center stage, with their pure white plumes standing erect. Confrontation is apparently not the point here, but an air of "Look at me, I want to be in the game" no doubt is.

Watching today for some observable resolution to the hooded merganser soapbox affairs has amounted to no resolution. I did see the birds catch and eat both tadpoles and fish. On three occasions, the tadpoles were three-year-old bullfrog adolescents. Hoodies are small ducks, so with large prey like those tads it takes a lot of whacking and thrashing to subdue them.

Once softened up, and probably either dead or unconscious, the tads are flipped into the hoodies' beaks and swallowed whole.

During an interval in the duck comedy of relations, a strikingly handsome cluster of eight common mergansers cruises by, searching for fish. They submerge their long bills and heads as they swim rapidly along. Such a picture! Like kids playing two games of tag in a schoolyard, these two merganser families share space and pursue prey independent of one another. For the fish, that is a separate problem.

APRIL 4

Watching hooded mergansers, for me, is like watching children on a playground. There is a lot of moving around, a lot of posturing, a great deal of switching from one place and back. With these ducks, it's impossible to figure out who is getting promoted to the honor of giving or having sex, and who is being left out. The bobbing of crests and the flying forth and back to the boxes continues.

APRIL 5

A development: A pair of wood ducks is positioned on the hoodies' nesting box this morning. They occupy the throne until nine o'clock, then leave. Two hoodies immediately notice and fly to the top of the box, only to be muscled off by a male wood duck who appears out of nowhere. No bird, to my knowledge, has yet entered the nesting cavity. What's next?

APRIL 7

The hooded merganser melee continues. One idea may have merit: Perhaps I'm seeing several immature males doing their best to become involved in this who-will-be-chosen circus. Duck follows duck to the box. At times, three fly up to the box, trying to pile on. This morning, a male wood duck flew directly into three others on the box and all were dislodged.

Two great blues today. Several common mergansers fishing, periscopes down.

APRIL 8

While the merganser game continues its ebbs and flows, I hear wild turkeys gobbling on the forest side of the house. Among the trees, heading for my driveway, are seventeen birds, five of which are males, each displaying explosions of feathers—visual proof of the power of hormones and the draw of a harem. Size must be part of sexual attractiveness, because as feather muscles stiffen, the whole turkey profile doubles in size. Most certainly impressive.

With males, there is little forward progress. Feet are programmed to take quick, short steps, never in a straight line, always edging sideways. As a result, each puffed-up male sidles as close to a female as possible. Primary wing feathers become instruments as their quill tips drag along the ground for the purpose of making scraping sounds. In leaves, there is only a rustling sound, but if the males can move the flock onto some hard surface, like my pebbly driveway, the scratching act goes into high gear.

Each male is the epitome of conceit as he struts his stuff. The desire to communicate "See me! See *me*!" is so intense that even bending down to eat is way beneath their dignity. This is narcissism on parade.

All the while, the ladies pay no attention whatsoever. Come to think of it, sex for turkeys—and for hoodies—seems a very complicated and frustrating business. In the midst of the overbearing, amorous advances of the gobblers, their harem oftentimes seems to become sick of the whole drill. They will, in their indifferent nonchalance, just wander off. It takes time for the bloated gobblers to even notice.

I usually have at least one surprise every day here at Great Meadow, and today the surprise comes as I watch what the hen turkeys do after leaving the party. Scooting away in some haste, they arrive at the edge of the Meadow. Instead of pausing when they reach the water, they wade right in, chest-deep! They begin to bathe like ducks: heads dipping, wings stirring up spray, all the while quick-stepping with their feet. As a result of their stamping, some type of presumed edible is stirred up and catches their attention. A lot of pecking at something in the water goes on.

Again, when time is spent with wildlife, it is a challenge to keep your brain in neutral. But neutral is no fun, so you try to be smart and predict. There is a trap there, because I am observing wild turkeys and they know what they are doing but I don't.

Here's what happens after the surprising swim and bath. About twenty minutes after leaving the company of all the pretty boys, back up the hill come all the ladies, bathed and spiffed up. When they get to the edge of the

Common Grackle

driveway, each lady begins to rapidly scratch in the soft, dusty dirt.

In seconds, the dust bath rituals begin. First, each female turkey deftly flips small chunks of dusty grit helter-skelter and skyward, using the tips of her primary wing feathers as shovels. As the sandy, gravelly particles filter through her feathers, her feet also shovel the loose dirt all around. There is a lot of kicking.

Following the initial dirt-shower comes phase two, the real dust bath. By now, enough excavating has been completed so that each bird can settle down into her own dirt tub, where she engages in the gyrations necessary to bring turkey feather-dusting to a climax.

Now, wouldn't a person normally think that a water shower should follow a dust bath, and not the other way around? That's the problem with making assumptions about wildlife.

And where have their consorts gone? Perhaps witnessing the after-bath toilette is considered improper by such gentlemen.

APRIL 9

Tree swallows. The branches of several bushy islands are festooned with these insect-eating transients.

A record: Four hooded mergansers on one nesting box! I try to ignore them.

A pileated woodpecker, a mustached male, is applying his superb grub-finding skills to an oak tree just beyond my window.

APRIL 10

J. J. Audubon's stylistic depiction of common grackles was based upon his observations of this impressive bird's body movements. The action of the male grackle, with his iridescent feathers, his aloof mannerisms, reminds me somehow of elegant women showing new garments on a fashion runway. With the exception of the bird's rather wobbly gait, grackles exude proper posture with every move. They raise their heads and necks slowly, always with a dignified grace.

When a grackle flies directly away from you, it turns its tail feathers to the vertical, using them to steer its flight as a sailor would use a rudder. The tips of the tail feathers are shaped like spoons, and the birds fit them together to form their unusual shape.

A grackle's eyes play an important role in its beauty. When a grackle directs its jet-black pupil, set diamond-like in his milky-white iris towards you, you notice.

The grackle's same-habitat companion at Great Meadow, the red-wing blackbird male, also possesses grackle-like mannerisms. When in the company of competing males, red-wings also cant their heads and beaks in high poses, but the movements seem awkward. Instead, the red-wing's banner of maleness is his brilliant, orangey-red epaulets. To display them most effectively, he must tilt his entire body forward. Once bowed, he extends his wrist feathers and flashes his colors.

Masculinity and sex appeal are clearly evident in both species.

APRIL 13

The sunrise this morning highlights gray, rounded, folded clouds moving westward. An occasional wisp of wind ripples the marsh water, turning the wavelets to silver. The weak sunlight shines at just the right angle to turn the islands of leatherleaf, with their still-clinging leaves of last summer, into patches of soft browns. It is oddly attractive how diffused sunlight creates reflection upon reflection, using only gray light.

APRIL 17

Spring peepers, more it seems than last year. American bitterns are taking turns at gargling, while pickerel frogs "gwonk" the nights through. Canada geese never seem to run out of gossip. Tree swallows continue to knit the air around the several boxes I've provided for nesting considerations.

Grackles and red-wings arrive to crunch sunflower seeds at the feeders. The grackle males are beginning their standard get-together with the boys, flying in gangs forth and back over the marsh. Spruce up, red-wing suitors: The first females have arrived to exercise their choices.

APRIL 18

So much can be learned by watching birds at feeders. Certainly there are similarities between grackles and red-wings, but when watching the two species feed on black oil sunflower seeds, the difference is stark. Grackles have jaw-crunching strength and make short work of the seeds' tough shells while at the feeders. Red-wings, like blue jays, must hold each seed under a toe and whack it repeatedly with their bills to gain the nut meat. That reveals something about their general diet as well.

Today's afternoon temperature has climbed past 80°F. The northern latitudes beckon to the common mergansers, and they comply. The past month's exciting observations of those special ducks ended this afternoon. Their dancing, their snorkeling, their fishing techniques, either singly or in cooperative flocks, their flashing beacons of white when they fly past with amazing speed—all will be missed.

One thing I wonder about: In all the time I've watched this species, never have I heard a single one of them utter a sound. May they carry the memory of Great Meadow with them and return.

This is the afternoon of the emerging turtles. Suddenly, painted turtles begin showing up on any elevated surface: eight on one muskrat lodge, fifteen on a floating log.

The snapping turtles have received the same wake-up call. Their carapaces cut blurry lines through the water as the burly animals reinvigorate their muscles and begin to search for food to end their five-month fast.

Osprey on the Perch Tree.

APRIL 19

Late this morning, the first sighting this year of a hooded merganser female actually entering a nesting box. A male swims around close to the shore. The female soon emerges from the box and flies down to the male. An egg laid?

This afternoon, three beavers huddle on a muskrat lodge. There is hardly room for them. Each is busy attending to itches that apparently built up over the long winter.

When they leave their lodge, its roof has been flattened. I doubt muskrats install roof braces for the likes of eighty pounds of roof-sitters. Muskrats seldom reuse a lodge. Unlike beavers, they don't use strong materials like sticks when building. This lodge will quickly decompose, and its original builders are probably already working on this year's housing.

APRIL 21

When the marsh air is crowded with a blizzard of tree swallows, there is not much hope for getting a lot accomplished. The show is just too intense and spectacular to miss. This afternoon, the birds have three focuses: the three nest boxes I've placed on small trees out in the marsh. The swarm of birds moves its attentions from one box to the second, then to the third.

They go in various patterns of flight. At times, the hundred or so birds fly in circles, some spinning clockwise, some countering. The next action may be a cooperative swoop, where the birds dive toward the nest box from all directions simultaneously.

Between the variations, they suddenly pinwheel downward together, landing on a convenient bush. A couple of minutes pass before the flock explodes upward again, this time performing their aerial stunts over a different box. With a hundred birds and three boxes, which pairs will win nests? Where will the others go?

At noon, a black-crested cormorant appears.

APRIL 22

A second cormorant.

APRIL 26

Two bouts of snow since my journal was last opened. It is probably just as well that words are insufficient to describe such white-sketched scenes, or that photography is equally inadequate. Eyes and an appreciative mind see the wonder of Great Meadow—its depth, its breadth, and its whole—and if the experience is rare enough, memory will retain it, without pictures or words.

Following a windy, cold night, the neighborhood's feathered crew has woken up early. I simply scatter seed over my entire deck now, and the place is abuzz with a spectrum of colors. The sheen of grackle feathers—blue changing to purple, then as the bird turns, hints of pink that flip to green, and somewhere within a feather comes a dash of orange. A strutting red-wing bows, calls "konk-a-lee," and flashes his shoulder calling cards across the deck. Confidence suddenly shows up: blue jay. Festooned in bold blues and whites, he exudes spirit, life at its most intense.

Diminutive avian visitors shuttle in, all hungry after a trying night. They come in fits and starts, some bearing caps of black, some exercising vocal chords that emit giggles, while other species land sporting saucy head tufts and belly-sides tinted with pale orange. Three mourning doves slant in, wings whistling, landing on the deck with pink feet. This choreography, this assemblage of life's gifts, gives the day a reasonably good start!

A truly majestic great blue heron

MAY 5

It is rather unusual, here in New Hampshire, to have an adult male blue heron reach the standard of feathered finery equaling those seen on birds farther south.

This bird stands high, his head crowned with streaming feathers long enough to reach the midpoint of his back. His long and willowy chest feathers fan across his torso and sway with the slightest movement of air. His epaulets, black and distinctive, top the wrist of each wing; from there, my eyes move down along the graceful layers of his primary wing feathers. Some finery.

JUNE 9

A soft, warm, summery world at Great Meadow. One might think that the sunset remembered the gentleness that persisted throughout yesterday, as its retiring rays follow suit by softening sharp-edged shadows.

A common snipe is calling in the dusk. Two beavers are slurping the young, tender leaves of pondweed, today's menu item of choice. A trio of ruby-throated hummingbirds takes turns sipping the sweets offered at the feeder.

I must take drastic action to assist the young plants just emerging in my garden. The peas have germinated nicely, up three inches now. At that size, they begin to reach out slender arms whose purpose is to grab the nearest object and hang on tight to support the plant as it grows. Tomorrow, I will put a fence in place and a tendril will find it and give it a hug. What stimuli in those tendrils give them direction?

JUNE 12

Early next morning I go down to the garden with a roll of wire and stakes under one arm, a sledgehammer in the other hand. . . . Disaster! Overnight, every green molecule of every pea plant entered some animal's stomach. All of those hoped-for tender delights of early summer are gone. Back to the shed I go, carrying equipment and dark thoughts.

JUNE 14

Last fall and all winter right through to spring drought held its heavy hand on the land. Finally, two weeks ago, helpful rains began to bring relief. Rain gauges measure around seven inches. One can imagine the Earth, the plants, the streams, and all flora and fauna sighing with relief.

The rains raised the water level in the marsh, and almost as if that change was a signal, the snapping turtle females are now on the move to locate nest sites. Over the past three days, trial pits have covered my driveway, none completed, none acceptable. The soil is too rocky, and this property owner feels guilty, as always.

JULY 20

Summer's calendar retells its past stories over, around, and in Great Meadow. Following the forgiving rains, the greening of the marsh surface proceeds rapidly. The prolific pondweed, its first growth tender and succulent and preferred by beavers, has now matured, blossomed, and toughened.

Algae masses are blowing gaseous bubbles; slow farts come to mind. The floating bladderwort plants, with their meat-eating vacuum bladders that suck in hapless zooplankton by the zillions, answer summer's call to procreate by sending up orchid-like blossoms of butter yellow.

The early summer music of tree frogs, while pleasing beyond words, has been far too brief this year. Their trilling lasted less than two weeks. The bass-playing bullfrogs are few in number compared to last year's chorus of boomers.

On the positive side, the bullfrog tadpole population is nothing short of astounding. Paddling the canoe through the gardens of pondweed, scattering the tadpoles, is like strolling in a summer hayfield where grasshoppers spring away by the hundreds. Here in the marsh, young frogs panic and splatter their way in all directions across the green sea of pads. Some dive below the green roof. Others skid to a stop in plain view and then panic all over again. How many will survive?

Every bullfrog female deposits some twenty-five thousand eggs in sheets over the water's surface. Estimates report that fewer than ten from each brood will reach adulthood. It would seem that those odds increased significantly last year. I must remember, though, that bullfrog tadpoles may spend three years as tads, and many mouths are waiting to keep the average in place.

JULY 24

Now, with fly rod in hand and my canoe as a platform, both sport and dinner

Deer snack at Great Meadow's table

exist right in my backyard! Yellow perch is the main draw and they have grown since the re-flooding to a perfect size: seven to eleven inches. We've had several meals already, for which I am grateful. No matter where I float my canoe, perch are there. One can only imagine the number supported throughout the entire marsh, and the number of perch probably pales when compared to the number of pond shiners and other fishes in the Meadow.

When I paddle my canoe out onto Great Meadow with a fly rod in its bow and a fine meal of yellow perch in mind, I become a predator. As a man, I'm conscious of the fact that I am taking a resource for pleasure. The natural predators who live in or around the Meadow or the ones to arrive from elsewhere are takers by necessity. Man's practice of taking has often extinguished wild resources. "How many are too many?" is a question that constantly needs to be asked.

JULY 27

The summer chorus continues, but with fewer voices. While last year competing groups boomed out their challenges and threats in vast numbers, this year I hear no more than a half-dozen bullfrogs. So the appearance of all those new tadpoles a few days ago is promising. I do hear green frogs, but only some. Pickerel frogs? I've yet to hear the grinding utterance of even one. Those sounds are missed. Great Meadow is not whole without them.

JULY 29

To add to my concern, Canada geese are teaching fewer offspring this year. I have only counted two families of black ducks. Today, one large brood of hooded mergansers was making a pathway through the pondweed and duckweed, the scramble of so many young proving to be uncountable.

AUGUST 25

Early this morning I spy a bald eagle on the Perch Tree. This adolescent, close to adulthood, is one large bird. Its mottled feathers, with many white ones scattered around its head and neck, tell its age to be near three years.

The bird's morning is spent preening and pooping. Around eleven o'clock, I watch the eagle spread its wings and lift easily into the breeze off its high perch. The day is warm and the eagle finds a convenient thermal, circling, circling, orienting onto whatever natural messages there are to be deciphered from the wind. The eagle seeks opportunities, analyzing them with the keen senses of its species.

My eyes follow until the bird's image is absorbed by altitude. Is it improper to think there was no other purpose but joy in that exercise?

AUGUST 27

Something new again on Great Meadow. In an earlier spring entry I noted with some surprise the appearance of a mother black duck with ducklings well before remnants of winter ice had melted from the marsh. It didn't seem possible. . . .

This morning, I see the reverse: A mother black duck closely followed by four small ducklings! Again, how can this be? So late. Have those tiny paddlers enough calendar days left to mature, to fly, to move south before ice time comes? This is the year of hatching black duck bookends, both pushing the limits.

AUGUST 28

Black duck family appears again this morning, giving pause all over again. Eat, learn, and grow little ones, and hurry.

For three days, a couple score of cedar waxwings have chosen to flit about the marsh with their distinctive helter-skelter fluttering. Watching them twist, turn, and bank during late summer sunset's light pleases me. At a distance, their tan breasts catch my eye as their wings bank their load in spiraling turns. At close range, when the angle of sunlight is just so, blinks of head-red and tail-white appear briefly. The freedom of migration is beginning to fill the airways over Great Meadow.

AUGUST 31

Bald eagle on the Perch Tree this morning; a mature bird this time.

SEPTEMBER 11

This morning, the female black duck floats by. Her brood, now minus one, totals three. Those appear to be about one-quarter grown, one quarter of their way through the race to beat the ice.

One must not pass this day, September 11, without pausing to remember the tragedy of one year ago. Bird songs, sunsets, a city park, a butterfly—not one element of nature's universe will ever be part of that tragedy. Great Meadow offers solace during my moments of reflection, though they are tinged by apprehension and I am full of questions.

SEPTEMBER 15

Before me I watch a patrol of hungry river otters terrorizing marsh fishes this morning. They aren't playing. They are showing the serious side of their natures—predators must eat. I understand what I am seeing.

Even under September's mat of aquatic greenery, the fishes, supposedly hidden among those thickets, are not safe from this twist of marauding otters. The adults, numbering maybe five or six, pour themselves over and under and through the pondweed jungle, and time after time, in intervals of only seconds, their whiskered faces emerge with jaws already crunching silvery fish.

It is obvious that otters lead dual lives. With this species, togetherness works both when bellies need filling and when the urge to play must be satisfied.

Cedar waxwings still sketch tinted lines over the marsh, intercepting insects.

Yesterday I welcomed the first snake I've seen on the house grounds: a fine common garter, female.

OCTOBER 5

While mostly rainless days perpetuate the region's dryness, beavers, following a summer of sloth, begin shifting gears to tend to their domestic duties: dam patching, lodge repair, and, of course, fattening up. Great Meadow's flat-tails are no exception, and while the water level in the marsh had dropped eighteen inches, with dam leaks plugged the water level is now static.

To date, fewer ducks have migrated through than during previous years. Only two sizeable flocks of Canadas (one on the water this morning) have dropped in for a night's respite.

I see bald eagles on most days. I see grebes and coots now and then, and otters continue their eating rampages.

The late-hatching black ducklings? I see the three remaining youngsters frequently and they are, on this date, mostly full grown!

OCTOBER 6

Duck hunters appear on Great Meadow. Shotgun volleys rattle the environs during early morning hours and again at evening tide, but from what I've been able to observe, either the gunners are inept or the ducks outsmart them.

OCTOBER 7

A magnificent mature bald eagle snatches some tidbit out of the water this morning, maneuvering the food from talon to beak several times while in flight. Eating on the fly, one might say. Within half an hour, the eagle selects two more courses of breakfast from the marsh menu.

Autumn's color show of leaves is just getting under way.

A northern harrier appears! The first of that increasingly rare species seen this year.

OCTOBER 10

The pattern of fewer flocks of geese and ducks continues during this migrating season at Great Meadow.

When the Canadas arrive, especially during the night, my heart quickens, my head fills with its usual questions. In the early morning light, my eyes strain to find their shapes upon the water, to distinguish their outlines from the hundreds of marshy islands across the expanse. How many in the flock? How far have they flown since their last rest? Have they been shot at and, if so, how many mates were lost? Might there be an injured bird in the group?

Come the first rays of morning light, the urge to move once again begins within the floating birds. Voices, silent through the night, now begin to speak, albeit softly. Wing flapping awakens muscles, stretching and readying them for the day. Some practiced order before flight comes to each bird, webbed feet responding and each body turning into the wind.

Like marathoners awaiting the starting gun, a similar stimulus touches every goose in this flock. Suddenly the air over the treetops is filled with the grace of geese. Usually flocks depart in groups—a few here, some more there—but not this group. These birds are all of the same mind. They follow a wide, arcing turn around the marsh edge before the skein works out its flying formation for the first leg of its journey. As the birds disappear southward over the treetops, I wish them safe travels and silently give thanks for the wonder of their being.

NOVEMBER 8

A day when the weather ranks a "10." Clouds reflecting sunrise-pink materialize across the ripple-free marsh. An hour later, the sun and most of its light is obliterated by a squall line from the west-northwest that brings a surprise shower of snow. Visibility over the marsh is reduced to waves of flying flakes that ride horizontally over the water. Flakes can defy gravity.

With the passing of the cold front, the sky clears and brilliant daylight reappears. The flakes have gone their way, but they leave behind a high, mischievous wind that yanks fall's spent leaves from their moorings. Now, instead of flinging snowflakes across the landscape, the wind abducts every leaf still clinging to its parent tree and gives it a final shaking.

Whitecaps push along the surface of Great Meadow amid swirls of water twisted by mini cells of wind, the turmoil creating conditions of superb joy for the several herring gulls here. The delight of these birds is obvious as they set wings at proper angles into the gusty wind, letting its currents blow them willy-nilly but never out of control. Gulls are masters at manipulating their streamlined bodies in gales, as they can turn wings sharply and tack at great speeds while swooping suddenly upward on one challenging gust after another. There they may stall on outstretched feathers, challenging the wind and managing to remain stationary and motionless.

My heart, laced with envy, soars with the birds. Joy-filled may not be the proper descriptor for these gulls, but as I watch them play with the wind, displaying their wondrous skills and enviable freedom, not unlike the eagle that soared in circles sunward a few weeks ago, what could be a better descriptor?

NOVEMBER 19

Two days of snow-spit and sleet, along with below-freezing surface temperatures, have left the marsh with a blotched, albeit handsome face this morning. Where the water depth is greatest, the surface remains open, but iced fingers grow ever outward, trying to seal it over. During this morning's sunrise, the frozen surfaces again reflect the soft pink of the southern clouds, the open water portions remaining black—a stunning contrast.

NOVEMBER 23

There is still avian wildlife to watch. Twenty-some hooded mergansers fish in first one open-water space, then another. Six black ducks rest in an adjoining ice-free area. I hope they are the same family that nested and brought forth ducklings on the Meadow so late in the year.

Today, I am surprised to see two Canada geese swimming and feeding together midway across the marsh. Apparently, they did not join those that left several days ago. The neck and head gyrations and verbal exchanges between the two are intense. One of the pair, I would guess the male, takes to the air, flying in a wide arc to the north, calling all the while.

He turns back towards his mate—but instead of landing, he flies low over her position. He repeats the circle once again, flying over his mate. The calling continues, but this time the male gains enough altitude to clear the tall trees to the south. He disappears down the valley.

The lone goose on the water continues to call, all the while twisting and turning in her place.

NOVEMBER 24

This morning, the single goose is swimming around with the remaining hooded mergansers, and she continues to exhibit her disturbed behavior.

An observer cannot help but ask questions. Late in the day, the mergansers leave. While understanding that nature lacks compassion, it is difficult not to feel some emotional connection to the goose. Her problem is unknown; but it is clear she cannot join her mate. What is known is that the ice will kill her, eventually.

DECEMBER 1

23°F. To date, the fall temperatures have managed to form only skims of ice across portions of the marsh. To assume that the freezing cycle of water bodies like Great Meadow follow an annual pattern would be incorrect. Marshes have warm spots, such as springs and areas of rapid decomposition, where water is slow to freeze.

The pattern of warm spots, as recalled from the winters of 2001 and 2002, in no way matches those of this 2002–2003 winter. In 2001 and 2002, long, narrow openings held iceless water, and the north end of the marsh stayed open for days after all other areas had been covered with ice. Only the area by the beaver lodge along the southern shore remained open after this year's freeze.

DECEMBER 10

By tossing scraps of this and leftovers of that over the marsh-facing deck railing, I receive wonderful rewards. By sunrise, the marsh's crows have awakened from wherever they spent their night. They are already perched high in the red oaks by the time I carry out my morning bird-feeding rituals. Naturally, when I step out onto the deck, the cautious crows scatter in cawing flight. Most mornings, I have snacks. This morning, the offering is leftover cat food and a few stale crackers, crackers that I painted with thin layers of peanut butter to add additional nutrients.

By placing crow offerings in a long-handled frying pan, I can slingshot the goodies over the marsh-side terraces and onto the steep bank beyond.

No sooner do I leave than from lookouts unknown come the spying crows. Then, my joy begins. Fly-bys begin at about treetop level, but with each lower pass, I know my black inspectors are planning their action.

Slipping on angled wings from high perches to low, individual birds become more brazen. The most courageous crows follow with close swoops over the scraps, egged on by a chorus of caws from the timid ones still high in the trees.

After those precautionary measures, one brave crow will bounce to a landing, hop one, two, three times, and make several fake grabs at a single morsel before finally making off with his prize. Now, with the danger proven false, that first bird is followed by a second and a third, and the goodies are gone in short order.

If the offering is something that takes time to eat, like slivers of meat on turkey bones, then the prizewinners grab the whole bone and fly some distance away to partake of their breakfast in relative peace.

If, as is the case this morning, the breakfast is light fare, then the gulp-and-swallow method is employed. On those mornings, one can usually look forward to a game of crow chasing crow. During that exciting time, this species' true mastery of flight is very evident. I automatically smile as every dive, spin, swoop, and fleeting curl is executed, as my admiration and respect for each of those black wonders manifests itself.

Fall settles on Great Meadow

2003

Pied-billed grebes coo. Killdeer cry.

JANUARY 18

To date, a winter worth a few words. There has been snow on the ground since early December. On Christmas Day twenty-four inches fell, followed by another foot during the next week. There was just enough moisture in those snows to accumulate on every cold surface and build up from there, crystal interlacing crystal.

The conifer branches bear the weight, their usual arcs of smiles bending into frowns. The bare branches of the deciduous trees have become tubes of white. Looking through such a maze, the scene seems dreamlike.

These days I put on snowshoes and am bedazzled by the white art. Every day I walk under snow-laden trees. No breeze disturbs a single twig. The low temperature ensures that the sculptures remain to defy description. I can't imagine missing such an opportunity. How fortunate I am.

JANUARY 19

This morning's temperature: -20°F.

Today's marsh and marsh-edge walk turns a few new pages. By following the tracks of three deer where they had plodded along through the deep snow, I am able to see how they vary their winter browsing. While on the marsh, they go from buttonbush island to buttonbush island. At each stop, they pluck the round seed buttons off each bush. When they mosey along the wetland's edge, they do not miss pulling off all the witch hazel seed clusters. I look into several of the two-seeded witch hazel pods and cannot find a single one that did not eject its pair of seeds last fall. Still, there must be nutrients in even those empty shells.

At midday, I just happen to be by a marsh-front window when hawk action explodes right before my startled eyes. A goshawk spins in a sharp turn near the top of a hemlock tree, folds his wings and plummets straight down through the tree's thick branches, driving his talons into a crow! At that same instant, the dozen crows left in the tree caw in fright and filter their way into the protecting forest, short one member of their clan.

JANUARY 20

For the third day, -20°F.

In the first dim light of dawn, a black shape appears far out on the snow-covered marsh. My binoculars reveal a lone eastern coyote.

Experiencing the moment when a wild canine clearly enters my field of vision is, for me, rewarding. These animals have become an extraordinary species. They now have their own taxon (a specific biological classification), *Canis latrans "var."* In a matter of only a short time, these canines, with a lineage derived from the smaller western coyote, have transitioned a couple thousand miles to the east. At some time during that movement, they acquired the genes and DNA of the gray wolf. That biological mixture makes it clear that a species, given the opportunity, does not require millenniums to undergo major physiological changes.

The remarkable animal before me, cautious, assessing, continues coming closer. It stands still and appears to be looking directly my way. I can sense that its inner caution light has turned from red to green as the coyote, stepping with what seems like unusual caution and purpose, heads straight for me. Its inner light turns red again and the animal stops. Like a sailboat tacking to fill a jib with some advantageous wind, the coyote strikes off at an angle toward the suspicious house on the hill, all the while keeping its head, with its keen alert systems, trained my way.

During this tacking trot, I can see this eastern coyote has a problem: an injured right front foot. Even so, it seems to lope easily on three good legs.

I expect what I see is deceiving. First, when one spots a lone and injured coyote, that may well mean it has been ostracized by its pack due to the injury, not being able to contribute to the group. Second, the life of a lone coyote is unnatural—and with difficult winter conditions combined with a major physical limitation, surely this animal faces an abbreviated future without a pack.

Still, the coyote's wild spirit rings bells deep inside me as an observer. The uncompromising hand of nature still reflects strength in the face of diversity as the animal lopes across the snow in the direction of its fate.

Chance may fill its stomach this morning, today, or tomorrow, but death is always the real master that awaits wild things that do not measure up.

FEBRUARY 20

Most of February has been extremely cold. With temperature readings in the minus tens and twenties day after day, cold gets into things. Days are warm if they reach 10°F. The snows since December keep piling up and never melting. Snow depths all around measure forty-five to fifty inches on the ground; a winter from the past.

While there has been no hard crusting of the snow, the simple depth of the never-melting cover makes it difficult for animals that must catch their food on foot, especially coyotes, foxes, fishers, bobcat, and deer. These animals are tough, however, and few except the injured and elderly will fail to benefit from spring's recuperative gifts.

FEBRUARY 28

Apparently, wild turkey scat in subzero temperatures is less than half its length and diameter than in other seasons. I thought ruffed grouse had left droppings in strange locations until I realized my error of identification. I learn every day.

MARCH 3

An odd date today: 03/03/03.

The tough winter continues. With last night's rain and sleet combined with this morning's temperature plunge to -16°F, conditions for wildlife are now serious. Deep snow, now glazed with ice, tests hoofed animals severely. Nature provides no alternatives. What is, is; and the challenge is now, a time that will separate strong from weak.

MARCH 17

Tracings of pink clouds viewed through a curtain of mist over the meadow; lovely. The light's softness courses through the greens of conifers along the far shore of Great Meadow. Only a temperature inversion above the cold snow produces this type of misty cloud, and when the backdrop is a sunrise-pink, these eyes feel the honor of the morning.

The privilege of sensing this morning's portent to spring is followed by the appearance of this year's first red-winged blackbirds. They number three, and indeed, seeing their black shininess is a three-fold thrill.

I cleaned the tree swallow and duck boxes yesterday. With at least two feet of snow still asleep on the land and across the marsh, snowshoes remain a must. All three duck boxes attracted egg layers last spring, but there were no successful hatchlings.

DUCK BOX COUNTS

Box 1: 23 eggs
Box 2: 8 eggs
Box 3: 3 eggs

The box closest to the shore has the highest count, the one farthest away the lowest. Neither fact probably has any meaning. I plan to put predator guards on each, as I saw a mink enter one duck box last year.

MARCH 20

Snowshoe walk today to the brook outlet of the Meadow. Crusty conditions remain, so tracking is very poor.

What is significant and exciting, and certainly worth a penned word, is my sighting of the year's first pair of Canada geese. They swung directly overhead, talking in clear voices over their wing beats. From spring to spring, the thrill never lessens; like greeting old friends.

MARCH 21

Two common merganser males took a tight turn 'round and over Great Meadow this morning. Finding no open water, they headed back toward the river to the south.

This afternoon, there's a bald eagle out on the ice, three-quarters distance across the marsh from my house. A large, mature male. He holds some morsel in his talons, and rips away bites with his beak.

MARCH 22

Same eagle this morning sitting on the Perch Tree. He spends the day flying to various perches of advantage around the marsh.

Otter out on the ice today. So good to have the promise of warm weather. Its touch brings wild things, with all of their fascinations and mostly secret lives. Better for all to stay that way.

MARCH 27

Eight to nine tree swallows make their swirling appearances over the marsh, in, around, and over the nesting boxes. Spring has arrived on wings.

MARCH 28

Phoebe! Seeing it back, like all the long-distance migrants, brings wonder to my heart.

The phoebe's southern home is somewhere roughly a thousand miles (or more) from where it stood on my deck railing this morning. The bird weighs barely an ounce, but what a package. Imagine the directional programming (memory, really) that keeps it on course, both going and returning.

I've removed last year's vermin-filled nest, and cleaned and readied the narrow shelf where a new nest will be built. Eggs will be laid, complete with mapping instructions for the phoebe's next generation of travels.

APRIL 1

Fool's Day

Today the countryside has been fooled by eight inches of snow. The storm has rudely quelled our short burst of spring fever. The weather stays very cold. The marsh's few open water slots have refrozen, and all is covered with snow again. The tree swallows and the phoebe have skedaddled somewhere.

Four deer nuzzling around near the house again last night, sniffing for acorns, having come every day since last month began.

APRIL 13

Warmer, with the marsh daring to show wider open smiles as its edges free up from ice.

Wood ducks cavort around the duck boxes. Nice to see the feathered royalty of those brilliantly decorated birds again.

Another thrill of this day comes when a pair of bald eagles displays their flight virtuosity, turning in tight circles just over the treetops between the house and the marsh. When their shadows pass, my mind reacts instantly, and I cannot help but stop and look up.

Following several minutes of that activity, their circling-in-concert reaches for elevation. They are only specks high to the south when they disappear from my sight.

APRIL 14

The female eagle (I believe) was scavenging on the ice early this morning.

Six or seven male common mergansers appear in the open water along the far shore. As always, they are a special form of life, so energetic, so distinctive, so good for the soul.

APRIL 15

Each day, a pair of osprey arrives. They, too, have ways of demanding attention. Often, they catch fish that are large enough to be difficult to lift up and out of the water. Their great wing muscles reveal their real power, when, with prey in tow, they strike the surface of the water and actually perform push-ups with their wings to gain the freedom of flight.

Peepers now spread hope-filled love notes throughout the early evening. The first voicing of the year from a pied-billed grebe came last night, and today, three painted turtles haul out of winter's chilly water in search of an early, warming spring ray from the sun.

A mature bald eagle presides over the Meadow this morning.

APRIL 25

Snow and cold yesterday but, even with the thermometer shivering at 23°F, the day dawns cloudless, leaving an uncluttered sky for the sun.

One would need twenty pairs of eyes to keep track of the bird activity on and over the marsh this morning. This year's flock of common mergansers, numbering fifteen to eighteen, raises the art of cooperative fishing to an astounding level. Oftentimes, when I focus my binoculars on those birds, I can see fish leaping out of the water in desperate attempts to escape.

Over the years, I have observed the great blue herons taking advantage of the opportunity of otters preying on fish. Now, the same scenario is being played out between the common mergansers and the herons. How many learned relationships among Great Meadow's wildlife have I failed to observe?

Literally hundreds of tree swallows dart about casting remarkable shadow patterns on the marsh surface. The colors in those shadows vary depending upon the angle and brightness of the sun.

APRIL 26

An osprey attempts to fish, but its breakfast is still swimming.

Again, wood ducks show interest in the nesting boxes. Where are the hooded mergansers this year?

APRIL 27

Swarms of tree swallows stir up Great Meadow's spring air. When an osprey surveys the scene from above and, uncharacteristically, sails off over the trees, something in my mind takes me back to days of Atlantic salmon fishing on New Hampshire's lakes with my father.

On whatever lake we were fishing, he was always very aware of a wide range of conditions that would either positively or negatively forecast the possibility of salmon for supper. For example, the direction of the wind was vital, as was the quality of the ripples on the water. Water temperature and cloud cover played roles, as did atmospheric pressure. When swallows flew close to the surface of the lake in search of insects, that was a sure sign that no salmon or lake or squaretail trout would bite any artificial fly we presented. His reason made sense: Fish are always wary of shadows. The swallows here at Great Meadow intentionally "throw" moving shapes into the water when they feed. That keeps the fish low in the water column and avoids any threat from above.

This morning, with the air above the marsh filled with shadow-throwing swallows, available prey fish are almost certainly lower in the water column. I'll wager the osprey recognized the fishing conditions were poor. I believe my dad was right.

Hooded mergansers are still late in returning.

APRIL 28

The common mergansers are stirring up a shallow area. Fish are panicking, and the action attracts a great blue heron. Less than a minute passes before the heron spots a fish and, with a lightening-fast strike of its beak, impales a foot-long horned pout.

This fish, also often called the common bullhead, belongs to the family of catfish, an interesting species because its members aren't like any other freshwater fishes—they have whiskers! Mostly, these blackish-gray fish feed at night. They use their sensitive whiskers, properly called *barbels*, to sense food in the muddy substrate of the quiet waters where they live.

Horned pouts have sharp "horns" or spiny fin bones—one on their dorsal fins and one on each of their pectoral fins—that can seriously puncture any soft tissue. The heron recognizes that it must exercise great care in consuming this family of fish. With the still-flapping pout secured in its beak, the bird flies from open water to the shore where the fish can be more easily controlled. Lucky for me, that spot happens to be directly in front of my viewing deck.

Repeatedly, the heron drops the pout and stares at it, beak poised. Any movement from the fish elicits a stab from the heron's stiletto beak. From my close observation point, I am surprised to learn that the beak is not clamped shut when it strikes the target, but is held open about an inch. With the prey impaled, the heron is able to lift the fish and shake it violently, without it slipping off.

It repeats the stab, drop, and watch-for-movement action again and again, aiming each spearing at the thickest portion of the fish's body where the majority of the vital organs would be. I expect the backbone does not escape stabbings either. My spear count from this predator is fifteen.

After the prey lays completely limp in the heron's beak for several seconds, the bird deftly flips the entire fish to point head-down into its throat and holds that pose for several seconds, knowing well the potential danger of the pout's three spiny fins. To avoid a fatal mistake, the heron manipulates the fish a bit along its conveyor-like tongue. The fish enters the bird's throat, but then is pushed back into the beak once again.

This trial swallowing occurs four times. Still holding the fish in its beak, the heron wades back into the water and dips the horned pout to wet it. Positioning the fish in its beak again, the bird tilts its head back and exercises its throat muscles. The horned pout creates a sizeable bulge in the heron's graceful neck, but slides downward into its stomach. The process ends with repeated rinsing of the heron's beak to clean it of scales.

MAY 5

Finally, a hooded merganser, a single female—cautiously, nervously, so it appears—shows interest in one of the duck boxes. Perhaps she has eggs in that box. If so, she has kept her actions secret from me.

MAY 6

I caught ten perch with my fly rod this morning, guaranteeing a fine supper.

MAY 14

My birthday! I wind my great-grandfather's old clock every week and simultaneously wind down my own weeks.

On the whole, I've counted far fewer ducks on the marsh this year. The pied-billed grebe continues to call, but its mating season has probably passed. Listening to those calls gives the word "plaintive" its perfect definition.

As the marsh continues another decade-long cycle of flooding caused by beavers, red-wings, and common grackles are severely challenged to find safe nesting sites amid the vegetation. With the water level at a historic high, the only nesting sites available are in sparse clumps of leatherleaf and within thickets of winterberry.

MAY 16

For over two years now, I've noted that both common grackles and red-wings have constructed traditional nests hidden within bushy clusters. Yet success in raising fledglings has been very low.

Great blue herons are the primary cause. With high water, the herons have few platforms from which to seek prey. Instead, they must land on top of the isolated islands and seek fish from there. The simple law of averages will result in a heron landing one day where a red-wing or a grackle nest has been secreted. No heron is going to pass up an easy meal of bird eggs or nestlings.

The result? Fewer and fewer black birds gracing this marshy world. When they are subtracted from the overall ambience of the marsh, much is lost.

Ah, but from absolute necessity often comes adaptation; the inherent capability of all successful life forms. From my canoe today, I discover a remarkable example in the form of a nest belonging to a pair of common grackles. This pair, instead of placing their nest in the crown of a cluster of winterberry, as is traditional, has built a nest close to the water, squeezing it in among the small trunks of the bush. In this non-traditional habitat, the pair of grackles has engineered an unconventional shelter deep in the center of the thicket.

To create a framework for their nest, which is supported by thin columns of living sticks, the grackles had to ignore any previous experience or inherited instructions. Instead of the usual biscuit-shaped affair, impossible among those stems, the nest had to be built long and cylindrical in form. The grackles used cattail leaves to form the basic scaffolding, which I'd never seen done before.

From rim edge to base, this nest measures roughly thirteen inches high! While the construction is ragged, the primary nest is a traditional cup, with a variety of colored grasses circling the interior. The cup is finished smooth with saliva-treated mud. Five bluish eggs, strikingly decorated with splashes of black, await the warmth of their creative parents.

MAY 17

Spied in mid-marsh between the Perch Tree and the far shore: our first Great Meadow moose! With my eyes practically stuck to my binoculars, I watch as the yearling moose becomes grace in motion, almost gliding through the muck of the marsh. With moose, the two words "gangling" and "graceful" form an unlikely pair of synonyms.

About thirty minutes later, while stepping out the door to bid goodbye to a guest, I see the moose standing not one hundred feet away, framed against the gray stones of the driveway. He lends us his brief attention, loses interest, and lopes into the woods, where shadows dissolve his form.

MAY 27

A pair of killdeer has returned to nest alongside the driveway, fifty feet from the house. Their simple, barely noticeable depression of a nest is totally open.

While the pair exchanges nest attendance duties, the male still makes advances upon his mate. She does not seem to object, and stands still while he struts to her side, fluffs his feathers, and neatly pops atop her. The pair remains stacked, statue-like, for fifteen to twenty seconds before he hops down; no mating action takes place. Afterward, one or the other returns to the nest and squats in it, as if trying it on for size.

One rather puzzling action practiced by both birds, more seriously by the female, is the persistent picking up and dropping of all sorts of detritus found lying around the nesting area. Pieces of bark, lengths of grass and twigs, small stones, all are pecked at, held briefly, then, with a shake,

Killdeer eggs are hard to see
Hint: There are four eggs

snapped away. This activity all takes place four or five feet from the nest. Camouflage à *la* randomness.

MAY 30

First bullfrog voices, and during the day.

JUNE 3

Though heritage inculcates a wide variety of camouflages in bird nest construction, there is no guarantee (especially with ground nesters like killdeer) that mere chance or unlucky circumstance won't lead to the loss of eggs.

This morning, the killdeer's four eggs are whole no more. From the size of teeth indentations in one of the empty shells, I would guess either a skunk or a raccoon happened upon that nest and filled up with killdeer embryos.

Those caring parents had been steadfast in their incubating duties through a very wet and very cold spring, oftentimes enduring days of pelting rain as they sat upon those eggs. After watching such determined attention to familial duties, it's natural to want success to follow, no matter the form of life.

JUNE 8

Most every day, I can see three beavers expanding their waist lines as they seem to combine breakfast, lunch, and dinner into one extended and much-relished salad. Common sense makes clear that beavers favor certain plants over others. Why? Probably because they taste better.

This long, cold, rainy spring has had severe consequences for aquatic birds. I've seen zero ducklings, zero goslings. Most unusual, but long spells of cold rain and lack of warmth always reduce the chances of survival for early spring families of wild things, whether feathered or furred. Any successful nesting this year will probably produce smaller broods.

JUNE 10

A female wood duck guides four ducklings, the first group of this stubborn, rainy, cold, uncooperative year.

JUNE 17

If one could watch this marsh all day, every day, what might be learned? Even during sporadic looks, I discover remarkable moments—like one this evening.

As observed, beavers are regular afternoon diners, continually stuffing tender pondweed into their faces. This afternoon, one of the beavers has company: a female wood duck. Showing not the slightest reluctance in approaching a fifty-pound floating furbearer, this mother duck boldly makes a major nuisance of herself by repeatedly pecking at the beaver's broad tail. Even more recklessly, she twice strikes the beaver's back. After each peck, the beaver paddles through the pondweeds a couple of lengths and stops, hoping to resume his private dining, undisturbed.

No peace. The mother wood duck quickly catches up and, like a total nag, begins the pestering yet again. After about ten minutes, the beaver has had enough. He dives and disappears someplace under islands of sheltering bushes, leaving behind one triumphant mother duck.

A thought about beavers during their winter: When their lodge is surrounded with thick ice, the family—a female, her mate, and come mid-February one or two kits—must remain inside, confined in a limited, dark space.

View of Great Meadow from the deck

One would think such inescapable family closeness would test togetherness to its limits. Apparently with beavers they simply accommodate.

JUNE 24

A large white-tailed deer materializes within the shadowy habitat beneath the crooked, ancient limbs of a marsh-edge winterberry bush. It is as if the afternoon light that had hidden her reddish-brown form simply slid away like a curtain, revealing the deer. There she is, sheer grace, smooth form, attentively stripping young winterberry leaves from their branches.

She is standing in a foot of water, and when she moves for a better angle at the browse, she doesn't simply shuffle her feet. She raises each hoof up above the surface and slowly slides it back into the water with no noticeable ripple.

Even though I am paying strict attention to this delightful action, the scene ends as it began; the deer seems to have been reabsorbed by the winterberry bush. There is no other way to explain her disappearance.

JUNE 28

In any forest, when spaces are opened to sunlight, there is always an army of eager seeds ready to move into that open space and germinate. To pre-

My old canoes, patiently waiting at the edge of Great Meadow

pare my view of Great Meadow when I designed our home, I carefully selected trees that could be harvested to open the desired view. If I had ignored the natural eagerness of the forest to replenish itself, my view of the Meadow would have been gone in less than four years. To prevent that, I keep cutting the tree saplings to allow low bushes and berries to cover the area. As intended, those plants have attracted a variety of browsers, and I'm pleased to discover that bush-nesting birds, like catbirds and rose-breasted grosbeaks, a song sparrow and a chestnut-sided warbler, have already built nests. I'm sure there are others.

A second mother wood duck, this one with a brood of eight tiny paddlers, shows herself this afternoon. The first mother of four ducklings now has but three.

JULY 3

During a midmorning canoe paddle, I chance upon what is not an everyday sight here in New Hampshire: A pied-billed grebe, a bird listed as threatened in this state, comes swimming out of a thicket of tall bluestem grass. Tucked behind her, scuttling erratically, comes three grebelets! Soon, a fourth emerges, paddling rapidly to catch up.

Within a couple of seconds, the adult notices my canoe. This female, like many species of waterfowl, deflates her swim bladder when she perceives a threat. The action allows her body to sink, leaving only her neck, head, and chunky bill showing. Her offspring take no notice and, threat or no threat, continue their investigations of everything in sight. The word "cute" doesn't measure up in this case.

While this family of birds parades by, the male grebe keeps "cow-cow-cowing," hidden somewhere beyond eyesight. So glad the marsh has become a refuge for this troubled species.

AUGUST 7

Once in a while, a scene in nature suddenly appears that causes an unexpected smile. As noted, this year has not been kind to many species of wildlife, especially nesting birds. Today, however, a single event rejuvenated my dwindling hope. There they came, perfectly lined up behind the mother, chugging along through a narrow opening between thick pondweeds, beaks-to-tails in a conga line: one, two, three, four, five, six, seven, eight, nine, ten, eleven, twelve, thirteen, fourteen, fifteen, *sixteen* mallard youngsters!

One could not draw a more perfect alignment. Such a family. What a scene. This parent's second brood, hatched despite long odds, certainly beat the forecasts by a wide margin.

AUGUST 10

If, when company arrives, they express interest in the wildlife action here at Great Meadow, I simply invite them outside onto my deck. There is always something of interest to share. This morning I see a young moose down close to the Meadow's edge, yanking and pulling clusters of grapes off a vine I planted three years ago.

I began the summer with one hummingbird feeder, and, as happened last year, two females and one male hummingbird came. Three competing hummers at one feeder guarantee a considerable amount of aerial jostling.

I added a second feeder this morning.

I'm not positive, but those rascals' actions lead me to believe that both females have built nests and hatched offspring. For several days now, six hummingbirds have been vying for sugar power at both feeders—feeders that require fill-ups every day. The wing-begging of three of the birds, along with an almost constant squeaking, indicates those are youngsters. Which nest produced which birds can't be determined. What is obvious is that being on the same deck with six diving, swooping, whirling, arcing, wing-fighting, hovering hummers, requires all my attention. Guests are on their own.

AUGUST 19

Again, a moose across Great Meadow is feeding on stems and leaves of white water lilies, the staple, perhaps, of a summer weight-gain diet. This is a young male, sporting antlers that appear to be about twenty inches long. His fur is darkest brown, and at a distance appears black.

AUGUST 20

Eagle on Perch Tree early this morning. I see it flying, and note that this bird will be very large come adulthood.

AUGUST 28

Two American bitterns fly in, only a minute apart. Each lands on a different grassy island and promptly takes advantage of handy camouflage. One, while difficult to pick out even with my 15x50 binoculars, disguises itself by pointing its beak skyward. In that frozen pose, its striped, slender neck blends in with the surrounding grasses. The second bird performs the same magic act and immediately becomes invisible. At this late date, I expect these two bitterns are migrants.

AUGUST 30

Both bitterns show themselves at different locations in the marsh this afternoon. I expect they are refueling.

SEPTEMBER 30

Once again, I witness the advantages of co-relationships and interrelationships among Great Meadow's dwellers. Like the great blue herons that score food by following the actions of river otters, in the past week I've seen a similar relationship among the mallards, black ducks, and pied-billed grebes.

As the ducks tip up their bodies and forage among the underwater weeds, usually in groups of four to a dozen, quite often one or more grebes swim and dive right along with them. I speculate that as the paddlers stir up the water world beneath them, fish, insect larva, and probably macro invertebrates are flushed out from hiding places and become food for the diving grebes.

OCTOBER 7

Great Meadow is active indeed: osprey yesterday, many ducks, several great blue herons, and four pied-billed grebes today. I saw a pair of river otters harassing and then eating scores of fish. Overall, a grand end of the year as the water level in the river, its brooks, and in Great Meadow remains high and productive for the late arrivals.

OCTOBER 14

At this time of year, the engine of photosynthesis in deciduous trees is running low on fuel; its supply of sun is diminishing. As the life-giving green fades, the underlying mineral players in each leaf are finally given the opportunity to show their stuff.

While the leaves were growing, minerals in the ground traveled up the pathways of roots, trunks, branches, and leaf stems to eventually dazzle the human world of leaf peepers. Once the spectators return home, so will the leaves they ogled. A breeze will set each leaf free, returning it to the ground to be recycled into nutrients for the coming year.

This fall, the show from valley to mountain summit has taken place in slow motion. Now the days are color-filled. The marsh explodes with tinted light, morning, noon, and near-night.

Small skeins of Canadas arrive and leave. Some land during late afternoon, rest, then take their leave in darkness, their plaintive "goodbyes" finding my ears as I loiter on the verge of sleep. Other unknown migrants settle in during the dark, rest, and then answer to the beckoning of migration as the sun awakens the marsh the next day.

OCTOBER 30

A pair of greater yellowlegs skips through this morning, feeding for only an hour before yodeling their way to their next, pre-mapped feeding station.

I see beavers, usually three, most every evening just before sunset. I suppose the number three means that last season's youngster will continue boarding with its parents in its birth lodge this coming winter.

Otters, up to a half dozen at a time, dive and munch ill-fated fish during both morning and sunset times.

The bronze days of late fall begin to arrive in muted glory; the richest of browns, the deepest of purples, the smooth variability of the tans. This second wave of performers can now burnish fall's stage with no visual distractions from brazen reds or bossy yellows.

OCTOBER 31

The month's final sunset is an eye-grabber. Rain fell from dark stratus clouds all day, but the front passed quickly, lifting the day's curtain to the west only minutes before the sun was ready to dip behind the watershed's high hills.

That brief opening has manufactured a spectacular sunset. Rain clouds scudding away to the east are infused with rays of Halloween orange. So much color splashes against that backdrop of clouds; great congregations of oranges ricochet down onto the inky waters of Great Meadow. The duality of that silent sunset luminescence sends me racing for my camera, knowing full well that no lens, no film coating, could ever capture the grandeur I have been witness to.

NOVEMBER 3

Just after nine o'clock, around one hundred Canadas turn in sync, making a wide arc over Great Meadow, and catching my eye as I sit in my office. Slowly, lowering altitude, they drop down onto the surface in three waves. Were they checking shorelines for the black metal of gun barrels before making the decision to coast in?

Each goose in these skeins seems larger than the summer residents, larger even than previous migrants I remember. They also act differently. Instead of clustering in a pod of birds and quickly going into the head-turned-onto-the-back-tucked-under-the-wrist-of-the-wing resting mode, these birds scatter in small flocks and explore most of the open water offered by the Meadow.

This grand group stays active, and they also stay silent until they leave, unseen, sometime before dawn. Had Great Meadow ever served these birds' families in the past? Did they take a memory of the Meadow when they left, a memory that might guide them back one day?

NOVEMBER 6

A flock of thirty-one mallards. Haven't seen anything like those numbers of migrants at Great Meadow before.

NOVEMBER 7

Eight ring-necked ducks staying very close together on the marsh. I don't see them even attempt to eat. Beak rings on three of the ducks stand out clearly, and one seems to have extra feathers on the back of his head like a crest.

A pied-billed grebe is giving them a hassle. The ring-necks, having flown who knows how many miles today, are trying to catch under-the-wing winks. As soon as they get settled, the lone grebe dives from some distance away and pops up directly between one or more of the visitors. Of course that causes each of the birds in the immediate vicinity to erupt from drowsiness and the commotion upsets all the others in turn. This game goes on for at least thirty minutes.

Meanwhile, an oscillating scrum of river otters is hunting for fish snacks. At least five make the water boil with their slick, glistening, diving, and curling bodies. Again, a great blue heron, recognizing the possibility that food may result from the otters' actions, lands close by.

Canadas land, rest, exchange verbal whatevers, and continue their journey every day. Sometimes there are eight or ten, sometimes between sixty and eighty. I hope for their safe journeys.

DECEMBER 2

First snow: fourteen inches. Canadas have been on the Meadow practically every night and day until now.

The white precipitation is nice, but couldn't it have waited until Great Meadow offered a smooth surface acceptable to steel blades? I just purchased a protective helmet, too, as a result of a bloody head-bash against the unforgiving ice last year. Somehow I had managed a half back flip and was knocked out for a while. The whack jumbled my brain enough to put my conscious mind to sleep, for how long I was unsure. When I stirred, the color red stained the ice and my vision refused to cooperate. Finally, I was able to remove my skates and climb the hill home. Needle and thread at the doc's stopped the leaking. Barbara was not pleased.

DECEMBER 9

Ah, crows. I've caught very few glimpses of the black crew for some time. Oh, now and then playful winds entice them to practice glissading on high currents, but they have always been far above the Meadow.

Now, come snow, where are they? No surprise: there they are, sitting around on high tree branches between the house and the Meadow, looking like finials on lamps. Of course, they remember last winter's daily handouts, and have no compunction about fixing a black-eyed stare into the private sanctuary my home. They know their presence will bring rewards, and I feel no anthropomorphic guilt in sensing that. I admire and respect the crows. They have a keen brain and I've watched it work.

DECEMBER 14

After the recent fourteen-inch snowfall, thoughts of putting blades on ice faded. Ah, but in this neck of the world, hope may be rekindled any given day when dealing with the vagaries of weather. Earlier in the week, an inch-plus of warm rain poured a lot of crystal-dissolving liquid into that mass of snow on the Meadow. Sure enough it turned colder, and today the marsh wears a cover of ice.

The voices of expanding ice having grumbled throughout the night; come morning I answer the call and walk for a while on the marsh's roof. The surface is lumpy and bumpy between clumps of leatherleaf and buttonbush growths, but there are acres of smooth ice in between. When time allows, I'll lace on skates, don my helmet, and become a marsh observer moving in a higher gear than usual.

DECEMBER 21

... and a bald eagle brings its spirit to Great Meadow on the solstice. White of head and tail, it flies around the edges of the ice-covered marsh and lands upon the Perch Tree, where it remains until light is lost.

DECEMBER 22

Such an honor to begin winter in the presence of a bald eagle.

The commotion of squirrels and assorted birds vying for positions around the bird feeders compel the eagle to come closer for a look. Even before the eagle lands on the first limb of a nearby oak, the threat of becoming a snack has been broadcast to all targets, sending feet and wings into escape mode. After a time, the predator returns to the Perch Tree, having gained no calories for its efforts.

Although I've been skating for the past two days, I'm reluctant to go today, as I would probably scare the eagle.

A green frog rests on Great Meadow's roof

2004

Beavers grunt. Canada geese chat.

JANUARY–FEBRUARY

Other than the dumping of snow on December 2 of last year, no accumulation of white during this winter season.

MARCH 1

Still no significant storms. The entire month has offered only moderate winter.

It's the first day after Leap Year, and there, cracking sunflower seeds out on the deck railing, is a red-winged blackbird. An early arrival indeed, but given that preferred nesting sites are available only to those birds already on the marsh, the rule must be first here, first served.

When we think about wildlife, especially about those animals that migrate long distances and are so vulnerable to anomalies in weather, we should have the insight to understand that the genes of species like the red-winged blackbird must adapt to the demands of climate. If a species' genes are incapable of change, the species will not fare well. Weather conditions and the availability of food will test those survival qualities.

MARCH 11

First Canadas. Two—one of last year's pairs, I like to believe—cruise by chatting, following a landing pattern set for a small stretch of ice-free water at the top of Great Meadow.

MARCH 27

Several dozen tree swallows and a mature bald eagle on the Perch Tree, each sending vibes of spring.

APRIL 4

Five inches of April showers. That load on Great Meadow has tested the old beaver dam beyond its limit. Built in 1997 and weakened by a similar rain event in 2001, the barrier of interwoven sticks, clods of mud, and roots have given way to the unruly, roiling forces of water that hustled the whole lot downstream.

APRIL 12

Now, at least three quarters of the water in the Meadow has gone to meet its mother river, the Piscataquog. The remaining water surface is constantly being stirred up by aggregations of common mergansers, wood ducks, and Canada geese. Blue Herons gorge on fishes stranded in puddles. My wife is concerned about the loss of our "meadow lake." Understandable, inasmuch as we now overlook acre after of acre of matted grasses, brown pondweed remnants, and mud.

I have a memory of the Meadow without water. For many years, I mentored students here when just a meandering brook provided the only surface water, a time when the Meadow offered acres of grasses, the ultimate in wetland protective cover for animals seeking similar habitat.

There is no question that the beavers will rebuild the breached dam. I hike over to see just what is left of the old structure after the floodwaters' ravaging. I can only shake my head. Where the old dam stood is the bed of a brook that's been scoured down to bedrock. On each side, only fractured sticks protrude from the ends of the old structure. It is going to be highly interesting to discover how the two beaver families, whose livelihoods depend upon Great Meadow, will handle the problem.

One might think the elder builders who designed the dam in the first place would undertake the construction. When I look at the gaping hole that must somehow be sealed by engineers whose construction machinery consists primarily of four orange incisors, I should be skeptical. I'm not. I've seen their ingenuity at work many times.

With the water volume severely reduced, Great Meadow's swarming fish populations have been forced into smaller and smaller quarters. Because the water level dropped so quickly, over a matter of three days, many fish were stranded. They did not go unnoticed. Herring gulls (a few of which frequent the Meadow on a regular basis), seeing the bounteous larder, apparently sent out word to their companions who make a living scavenging at a city landfill some twenty miles away. They responded in crowds and found fast food at this inland place. Every day, all day, the air over Great Meadow virtually sparkles with the twisting, swirling, slanting wings of white gulls. The action is accompanied by voices of both goose and gull.

APRIL 14

Yesterday, two ospreys came to fish and eat but the Meadow's resident overseers, the crows, rudely interrupted their dining.

The Perch Tree supported the pair of ospreys, both trying to dine on their catch. A couple of crows, always seeking some diversion, proceeded to harass the ospreys. First, they sat on dead limbs directly above each diner and tried their traditional "caw-them-away" confrontation technique. The ospreys took notice by repeatedly turning their sharp-beaked heads and glaring at their tormentors, but it was obvious the gang members weren't about to be deterred. Despite all the harassment, both ospreys went about gouging large chunks of flesh from their prey in their usual manner of indelicate dining.

The crows, however, being crows, stepped up the fun. It was apparent that if their in-your-face cawing wasn't sufficient, additional bold action was called for. First, one crow tipped off his perch and actually landed upon the back of the osprey perched directly below. That was too much and the osprey, with fish in claw, left the perch. Only seconds later, the second crow dislodged the second target in the same manner.

Although the first osprey disappeared down the brook valley, the second circled, still clinging to his fish, and returned to the Perch Tree. He was there only a very short minute before he was accosted once again, this time by both of the black pranksters. This time, having had enough, he flew off with the raucous mocking "caw-caws" of his tormentors following his retreat.

Heard the call of a killdeer today. Will they nest once again close by?

APRIL 17

In late afternoon, four ospreys appear over Great Meadow. All four birds, each with a fish, bring grand activity to the old Perch Tree. Today, the crows are apparently attending to business elsewhere.

The quartet of ospreys dispatches their small fish quickly and soon after I see all four of those splendid, long-winged predators hovering in the air. With the Meadow's fishes now crowded into small pools, hunting is good. Watching the diving action of this small flock of birds is even better. After each plunge, after talons have exercised their purpose, after each bird catapults up and off the water, they shake away the heavy loads of droplets in a visible spray. Never does that action, peculiar to ospreys, fail to please the eye.

Now comes another reason to admire this species: An osprey's wrist structure allows it to actually swivel the head of every fish caught into its line of flight. Because the body of a fish is streamlined, air resistance is minimal. Flying with a sizeable fish is not a problem, but landing on a perch with one taloned foot while trying to control a still-struggling fish with the other can be. On more than one occasion, I've seen ospreys lose their balance, fly off with their catch, circle, and try again. I watch intently.

While the fish struggles, the osprey somehow maintains its balance by flapping its wings. It becomes critically important to subdue the uncooperative fish as soon as possible. The osprey first attacks the fish's eyes and the tough bony structure of the head. Each jab of the bird's sharp, hooked beak is accompanied by the fish's violent, muscular jerks. The bird finally subdues its supper, savoring it for the better part of an hour.

Were these four birds single hunters or two pairs that came together on this particular afternoon at a good fishing hole? What really mattered was that their sixty minutes of riveting entertainment gave me a proper excuse to linger and enjoy my martini.

APRIL 19

What with the plethora of gulls, common mergansers, marauding crows, and literally squadrons of tree swallows, plus the ospreys, any observer, no matter how nature-challenged he or she might be, cannot help but feel some of the wonder of the marsh on this spring day.

Nor could they not notice how quickly the activity changes when a bald eagle comes to visit. Within seconds, its presence sends warning peals throughout the marsh: "Who wants to get eaten?"

APRIL 20

American bittern flying across the length of the water-depleted Meadow.

The very active common mergansers keep Great Meadow's now scant surface water rippling from their ceaseless diving activities.

Tree swallows buzz by the dozens now, around and around each birdhouse. Who wins those offered sanctuaries? And by what rules? My questions are always the same, but the birds work things out morning after morning, with a single female winning the nesting spot prize. Once she is seated, most of the throng moves on, but tree swallows are seldom without company.

APRIL 26

The ingenuity of Great Meadow's beavers has been proven yet again. Surprise? Not really. Water levels have begun to rise, so a walk to the site of the Meadow's blown-out dam is in order. Once there, I simply stand and marvel.

Somehow, those master stream-tamers have figured a way to construct the base of their replacement dam on nothing but smooth bedrock. They first had to secure a series of long pole-sticks, fastening each pole with one end placed against stones several feet downstream. Instead of placing the sticks crib-like across the stream current, they did the exact opposite, aiming each pole directly into the current!

Then they had filled the cavities between the poles with root-bound clods of sod. From that clever foundation, they could weave a series of shorter sticks through and against the clods, which were held firm by those first foundation sticks. Once the base work was reinforced by water pressure, the dam could take on width and height.

Today, the crest of the re-engineered dam measures five feet above the bedrock base. Great Meadow is refilling. Normalcy is returning.

An immature bald eagle shows its inexperience by repeatedly diving into rafts of common mergansers. A losing game for the eagle, but fun to watch.

In the meantime, the rising water level provides new flotation for dozens of ducks and geese, and new fishing grounds for the ever-present great blue herons. The consequences of April's flood have been erased by the adaptive capabilities of the marsh's primary residents. They lost their watery world, rebuilt it, and in the process reinstated habitat for the wetland's conservatory of life.

MAY 8

First bullfrog "harrumph."

MAY 21

Meadow level back to full. No way to judge how the lowering of the water in April has affected the seasonal cycles of the Meadow's dependents.

It seems that the voices from spring peepers far outnumber those of past choruses, but I tend to think that every year. After ten days of mid-May cold and long days of hard rain, a short spell of warmth has stimulated the gray tree frogs, and they sing me to sleep during a heartening week of wonderful tunes.

A pied-billed grebe broadcasts "I am here" calls off and on through the day and night.

I welcome back nearly twenty Canadas. They engage in loud discussions all night long, the noise level at times building until the word "harangue" comes to mind.

The season of green and music has returned.

JUNE 25

How I hate to notice the rapidly declining voices in what has seemed to be the most song-filled spring since we came to Great Meadow: dozens of gray tree frogs trilling their pleading hearts out, hundreds of peepers clinging to grass stems along the Meadow's edges, competing with their tree-bound cousins. Together they've created bedtime music for several nights, but apparently the sexual frivolity had now passed as last night but a single gray tree frog's call reached my bedroom. A scattered few bullfrogs still croak intermittent stanzas, but their ardor seems to be weakening as well.

> **SIGNS OF LIFE'S CYCLE OVER THE PAST WEEK:**
>
> 7 black duck ducklings
> 3 Canada goose goslings
> 5 mallard ducklings

Goslings are few and the reasons unclear, as there were more geese on the Meadow this year than in the past. After reading Bernd Heinrich's

Deer munching on algae some distance away

The Geese of Beaver Bog, I fear it may be that the water's new height may have made private nesting sites hard to come by, or that previous sites were flooded. Female black ducks and mallards parade ducklings now and then, with two new broods just within the week.

JUNE 29

I spot a Blanding's turtle today, a large female trying to dig a nest cavity in the impossible stones of my yard. Satisfying to know that Great Meadow is accommodating a species of special concern in New Hampshire, but deep grief sets in as I watch this precious turtle, and others, search and dig in vain.

JUNE 30

Bright, clear day.

At the wetland's edge this evening, a deer stands belly-deep in the marsh, just off shore. It's a full-grown doe, sporting a reddish, velvet-like coat of the color so often seen in early summer. I watch with binoculars, fascinated, as she feeds from Great Meadow's table. Like a moose browsing for water lily tubers, this doe moves her submerged muzzle around in tight circles. When she lifts her head, long streamers of algae (*Chlorella* or *Spir-*

ogyra, I expect) drizzle in strings from her mouth. She catches those slippery strands of green with her gyrating tongue before they can slip away.

Each dip into the water brings up globs of nutritious algae that the deer seems to inhale, quart after quart. Her muzzle, from nose to just under her eye sockets, is layered with a film of green. After several dips, she extends her long tongue, and this time slaps at the strings of slime around her face until they're cleaned away.

She feeds for about fifteen minutes and averages a mass of algae roughly every forty-five seconds. When she paddles out of her aquatic kitchen, she enters the tall blue-joint grass and slips away into the forest. Who knew of such breakfast fare?

JULY 4

It's after midnight, early on Independence Day. Some innovators bent on merrymaking have mounted a train whistle on a pickup truck and suddenly it seems like trains are invading the surrounds of Great Meadow. The whistle, certainly authentic, fills the night. Indeed, being of an age at which only memory retains long-forgotten everyday sounds, I smile with nostalgia as I lie in bed listening to yesterday's echoes.

Barbara paddling through Great Meadow

Daylight brings a fine day to enjoy free time with Barbara. While appreciating waffles on the porch (accompanied by this year's crop of kamikaze hummingbirds, all female and numbering five), we agree on yellow perch for the evening's meal.

Once we launch the canoe, a bullfrog booms forth some pronouncement; a welcome, perhaps, to our first Meadow float of the year? We paddle our way through green algae mats, many puffed into gassy bubble formations, the result of decomposition and growth. Frogs, green and bull, leap for their lives as their pad-covered haunts are threatened by our perceived floating menace.

Just beyond the marsh's plant zone, a female moose, some two hundred yards distant, peers around a large, thick, winterberry bush island. Neither of us noticed her great bulk at first, as she had been neck-deep in water, and only seconds passed before she thrust her head forward and into plain sight; a surprise to us, and a surprise for her.

It's then that she does something that seems strange: She thrusts her long muzzle into the thicket of bushes. Our "why?" is soon answered. Apparently, upon our approach in the canoe she had directed her calf to hide in the cover of plants. We inadvertently created a threat too severe to ignore.

Now, the mother gives a distinct call, similar to a coughing sound, and the calf thrashes out of cover into plain sight. While the mother can walk in the deep water, the calf has to follow by moose-paddle. We watch, enthralled, as the pair make their way to the shore of the woods one hundred feet away. When the female moose reaches land and turns her head our way, perhaps it is best we cannot interpret her thoughts.

Ah, the yellow perch. Who can better such a supper sitting overlooking Great Meadow, where summertime offers fresh fish and just-picked peas?

JULY 13

A pied-billed grebe calls. I see it twice, both times swimming partly submerged, showing only its neck and head; there's some unknown threat, I'm sure. Might this one be a female?

AUGUST 31

Another grand sunset at Great Meadow. As I step out onto the overlook deck, the noise of the screen door closing flushes dozens of waterfowl.

Ring-necked ducks again: fifty or sixty circling, circling, landing for quick splash-ups, then airborne again, flying in no particular formation yet impressing me with their slanting, wide, colorful turns. They shift into overdrive and sail up-meadow.

Surprisingly, flying with them and keeping up the pace are three semi-palmated sandpipers. They break away from the racing ring-necks on occasion, but then, with a shortcut, rejoin the ducks in a remarkable series of tight-formation maneuvers. No problem with instantaneous cross-communication between these two species while flying at high speed. We humans know so little about the lives of other species yet often don't give those lives a second thought.

Meantime, as a mature bald eagle rips apart his catch on the Perch Tree, an unusual scene takes place mid-marsh on a buttonbush island. Similar to the nightly roosting of migrating herons on islands along southern seacoast locales, here are seven great blue herons on a single bushy island at Great Meadow, fussing to gain proper footing before settling in for the night. This is the first time I've observed this behavior on the Meadow.

SEPTEMBER 2

A sticky, humid air mass has just been dismissed by a late August flush of clear, cool air.

With the air now crystalline, the trees along the Meadow's far shore indicate their summer work is coming to an end. The greens of chlorophyll have begun to thin. Discernable today, but only in the subtlest of tints, are the colors of the underlying minerals that have sustained the working leaves during the sun-eating days of summer. Nothing transitions as beautifully into a winter's rest as deciduous trees.

SEPTEMBER 16

Hummingbirds left yesterday. Sometime during late August, the five females of the season dwindled to two. The destination of these remarkable miniature birds is either the panhandle of Florida or Mexico's Yucatan, a migratory feat that merits thought. Three major hurricanes are currently being tracked, with Ivan's fury sweeping into Georgia today. Such breezes treat flying fluff badly. I can only wish the hummers well, and if fate allows, maybe at least one of "my gals" will return next year with a beau in tow.

Due, I think, to a wet spring and summer, foliage is only now beginning to seriously give up its green. Just a few migrants rest at Great Meadow. Ring-necked ducks came again today, only to be repeatedly provoked by an osprey that seems intent upon hovering nowhere else but directly over them. Is there a purpose?

OCTOBER 22

Days gallop past in a blur. October's fine weather seems to be in a hurry to leave. Great Meadow's fall show is exercising its final act. Eyes must experience what words struggle to describe: leaves blaze and fade; angled light clarifies and amplifies every autumnal color. The wonders of fall migration arrive at the marsh—ducks by day, geese and mostly unseen songbirds by night—to replenish the nutrients lost during long-distance flights among the clouds.

On occasion, pellets propelled by exploding gunpowder strike their target and a feathered life crumples into Great Meadow. I, too, have pulled triggers and seen such birds tumble to their final splash. "Good shot," some companion hunter would say. While it is difficult to watch waterfowl tumble limp from the sky (especially those that have sought refuge in "your" Meadow), I hold no rancor towards the gunmen. If they are true sportsmen, age may well shift their mindsets from killing to appreciating.

NOVEMBER 18

On this night, we have just snapped off the bedtime reading lights and not a minute passes before the first "yip, yip" finds its way across Great Meadow to our ears. We both spring out of bed and bolt onto our bedroom porch.

Across the Meadow, in the lower field of a pasture, comes the sound of singing eastern coyotes! Their chorus enchants the night and we listen, spellbound. The coyotes' introductory, plaintive "yip" notes quickly morph into an elaborate flourish of canine harmony. I've heard this species' vocalizations many times here and there, even combined barks and mini group howls, but this distinct concert lasts over a minute, filling the Meadow with spellbinding music.

How many eastern coyotes in that pack? Why that sudden expression of soaring yelps? This family is not stationary. They yip and bark and howl as they run through the pasture, across the marsh. As we stand bewitched, a final voice, a long, sustained, wolf-rooted howl, commands that the revelry stop.

The quiet of the night has been restored, but the audience in attendance remains sleepless for a long while, thinking about the wild conversation we experienced during that howling session.

NOVEMBER 21

It has been cold enough for ice to skim the marsh surface during the past week, but the sun's rays change those molecules into liquid today. Canadas remain. I expect the same dozen will stay on the marsh as long as they can reach enough nutrients. Otters fish each morning. Beavers feed each evening.

DECEMBER 1

Visible, but barely, a mature bald eagle endures a morning's downpour on a branch at the top of the Perch Tree. Not since midsummer has an eagle's aura spread over Great Meadow. Why visit today? The rain is pelting. There are countless perches in large conifers close by, where protection from the storm could be easily found. But there this grand bird perches, hunched against the wet, head swiveling right, left, alert in the drenching. Fog settles, sleeps, and then begins to rise again.

By late afternoon some clouds have unraveled to the west, and a strong beam of sunlight scatters across the Meadow. The light shoots into the sleepy fog like a laser, dispersing the vapor droplets in a giant orange firework.

With orange mist still in the air, a rainbow drops its flavors of light directly onto the Perch Tree where the eagle still rests. What an instant to capture on film—but the scene has struck my legs useless. A rainbow to my right, a shaft of orange sunbeam above the marsh; both cast brilliant reflections onto the black water.

Clouds to the west then shut off the ray of light, and the gray fog settles back into place. I look for the eagle. It has been taken someplace by the rainbow.

DECEMBER 31

Again, it's time to toss another calendar into the trash. If only it were so easy to forget I've lost another year.

Yesterday was my first walk on this winter's version of a roof over Great Meadow. The frozen surface provided a perfect medium for reading the foot messages left behind by last night's creatures. Evening forays in the snow leave clues for any skilled reader of tracks to decipher.

The trudging route of our resident porcupine was clearly visible. Porcupines often pee as they make their way to various eating places, leaving dribbles of urine as scent markers. I might, if I could, recommend caution to the wanderer, as its nemesis, the fisher, might be hungry. Just over the wall, a mink left hop, hop, hop tracks, going from the back of the wall, scaling a stone to its top, slipping off to the opposite side, and hop, hopping under my overturned canoe. It looped under the bow, belly-skidding in the snow, then hopped back to search for prey among the stones in the wall. Under low bushes along the marsh shore, a long-tailed weasel had picked its uneven way, probably seeking a snack of anything warm. A pair of fox tracks zigzagged then straight-lined their way along, following a map set by their noses.

More tracks also made clear that a pair of eastern coyotes took advantage of the brilliant marsh moonlight in their own quest to find a midnight snack. I saw that a bobcat must have cast its moonlit shadow during the night, and four or five deer took an evening stroll atop Great Meadow as 2004's curtain came down for the last time.

2005

Hummingbirds squeak. Coyotes croon.

FEBRUARY 11

Any time I walk past the many windows in my home that face Great Meadow, I glance out over the expanse. I do the same on the opposite side of the house, where thick forest dominates. So accustomed are my eyes to the views that any change is quickly noticed: a dark spot, a dead branch, a stick not there before. But late this afternoon, a truly wild sight sears itself into my eyes. There, statue-still, not one hundred feet away, stands a male eastern coyote.

The animal clearly shows himself to be at the physical apex of his latest evolutionary development. This eastern coyote embodies the meaning of the phrase "newly magnificent" in every physical trait: large head, with heavily furred and rounded ears, an actual mane covering his neck and shoulders, augmented with thick, light-colored hair along the spine of his back. Very long legs and a tail thick, full, resplendent.

He stands stock-still. No motion, not even of an ear. Assessments are being made.

He turns towards the marsh, moving in long, even strides. Then, he stops. He has caught a scent, a message of the kind that resonates to canines of most stripes: "STRONG ODOR HERE." Apparently the coyote finds it impossible to resist the scent, so there, not thirty feet from where I am watching, he accepts the invitation and rolls repeatedly in some odorous perfume. He stands, gives a satisfying shake, turns, and resumes an easy gait downhill, disappearing into a dark thicket of eastern hemlocks.

I hope this animal will cross out onto the snow-covered surface so I can catch sight of him again. Several minutes pass. No coyote. But then, far across and along the shore of the Meadow's southern peninsula, there he is. Still moving easily, about one hundred yards opposite the point of land, he stops, turns his head. At that moment, what must be his mate steps out of the forest and onto the open marsh, and trots to his side.

This female had apparently been waiting for her partner to pick her up so they could continue their rounds together. I watch as they trot in line to the far beaver lodge. They both investigate, looking for small mammals that often occupy such lodges. The female lingers. The male waits. When her curiosity is satisfied, she joins her mate and continues on. My emotions go with them.

FEBRUARY 20

The solo behavior of a small porcupine who lives in a terrace near the house—we call him Runty—has changed today.

It's late afternoon and high in a nearby red oak, Runty is joined by another of his kind. Is this a chance encounter? A purposeful one? Hard to read, but their body language keeps my eyes glued to my binoculars. Where a large limb joins the tree trunk, the two animals engage in a call to action where head-butting seems to be the program. Their wicked incisors seem to play no role in the activity, but their noses take some abuse after repeated feints change to serious bumping.

Is this sexual foreplay? A stay-out-of-my-territory bullying match? The much larger porcupine seems to occupy the high ground on the thick limb. After several minutes of the set-to, Runty tries a new tactic. He turns completely around, backs up, and slams his quill-studded tail into the opponent's face. Three quills find their mark.

Runty turns around again and the head butting resumes. The confrontation lasts only three butts' worth before Runty leaves the skirmish and backs down the oak. Once on the ground, he picks up a leaf-covered twig of favored forage and nibbles away in obvious contentment. Runty has left his calling quills to be remembered by.

Later, near dusk, Runty draws my attention again. I'm convinced (or better stated, I've convinced myself) that to our largest tree eaters here at Great Meadow, namely beavers and porcupines, trees of the same species do not necessarily taste the same.

Both beavers and porcupines consume bark from certain trees, with beavers preferring poplar and porcupines regularly choosing eastern hemlocks. Porcupines gnaw at and eat the outer, inner, and cambium layers. Beavers do the same.

In those hemlock groves closest to a den of porcupines, one will find certain hemlocks among the many that have been nibbled and gnawed to the point where only snippets of needles or branches remain on an entire tree. Hemlocks growing only a few feet away are left untouched, except for perhaps a few gnaw (taste) marks around the trunks' bases. Such trees are never climbed. I believe they are perceived to have a bad taste.

I suppose it stands to reason. Each tree's nutrient-gathering roots extend into soils that contain different mixtures of basic elements. Those mixtures are bound to flavor the portions eaten by porcupines, who have very particular tastes and consume a wide range of vegetation. They know good eating when they taste it.

Where beavers are active, trees are also taste-sampled. When a tree passes the taste test, the beavers fell it and consume every reachable, edible surface of bark.

FEBRUARY 27

Back to Runty. He has found several "delicious" hemlocks between the marsh and my home. Only skeletons of branches remain on the chosen trees. Close to the top of one of them, a small platform of debris has accumulated where several bare branches come together. It's getting dark, but I keep my eyes focused on the scene. This is strange. This particular platform is made of wide-blade grasses, like those forming the graceful leaves of cattails. There is a double mystery here: Who built the high nest? And why would a porcupine spend time in such a lofty staging?

Runty tramps around on that swaying mattress of leaves, apparently patting down the grasses with his feet. When he stops, he whacks the outside edges of the platform with his tail. (I often do the same to soften up my pillow.) This action is followed by what certainly appears to be . . . a nap!

On this night, I watch until my binoculars are no longer any help. Runty seems to be fast asleep on his high stage.

MARCH 16

Red-winged blackbird.

MARCH 28

Common grackle.

MARCH 31

Immature bald eagle on the Perch Tree. Several hundred yards to the north, an otter appears on the blackening ice, crunching a fish. The activity attracts the young eagle's curiosity.

Dropping from its overlook, the bird sails to a landing, perhaps fifty feet from the feeding otter. The reaction of the otter, only a medium-sized one at that, is surprising. It totally ignores the eagle, even going as far as to turn its back to the threat, hardly missing a bite from its meal.

The eagle, setting talon-spurred feet one in front of the other, marches toward the otter in a typical swaying gait. The bird swaggers eight or ten steps, then stops. Another few steps, then stop. The otter, either a brain-lite animal or one molded from the devil-may-care group (more likely), completely spurns the foe by rolling in the slushy ice, preening, scratching, and looking off elsewhere.

The eagle takes a few more steps. Now perhaps only twenty feet separate the two, who seem to be testing one another's resolve. Suddenly the otter settles the matter by sledding directly through the soft ice surface towards the eagle. The young bird immediately shies away, opens its grand wings and flies off, perhaps wisely. To sink talons into an otter would, without a doubt, set off an explosive reaction of sharp teeth and ripping claws, all commanded by a writhing pretzel of wiry muscle.

APRIL 5

A half dozen common mergansers, males only, along with a score of tree swallows. Ospreys are fishing late in the day.

APRIL 24

Uncountable numbers of tree swallows; as usual, they summon my watching.

APRIL 25

Of the four ospreys that fished at Great Meadow yesterday, one remains on the Perch Tree this morning. I expect that bird was watching, as I had been, the aerial show put on yesterday by the sudden congregation of hundreds (three hundred? four hundred?) of tree swallows.

Prior to this morning, flocks of this size only stopped here during migration. (I tried to focus on a single swallow and couldn't; impossible, like watching a single snowflake.) The birds would fly in a tight formation, then suddenly split apart, one cluster of birds wheeling one way over the Meadow and the other wheeling the opposite way, only to quickly join ranks again and plummet downward, barely skimming the water, allowing their shadows to push them upward into a fountain-like spray.

MAY 9

Ah, a new insight into ospreys.

Earlier, an osprey snagged a sizeable fish. It spent an hour pulling off all the chunks of fish with its beak—so, naturally, pads, toes, and talons were covered with fish oils and scales. How to clean them? Today I learned the answer. To rid its feet of the residue, the bird simply flew off, extended its legs, and dragged both feet along the water, shaking them over and over again! Being on hand to watch such a scrub-up was fortunate indeed.

MAY 14

Another day, even a birthday, is always a gift, something I'm thankful for. Indeed, this day arrived with a special reward from Great Meadow. Just before six o'clock, between emerging small islands of aquatic greenery, came tiny ripples—ripples created by five goslings under close supervision by their parents. The yellow, puffy ones are now busily snapping at tender green shoots of grass.

So the day passes. There is land protection correspondence to be done, but always sporadically, due to distractions from the Meadow; it's forever a juggling act here.

I watch the family of five yellow paddlers returning this evening from a day of "How to Become Proper Goslings."

Barbara says, "They're bigger than they were this morning!"

Yes, it certainly appears that way.

MAY 18

A killdeer makes an appearance just south of the house, perhaps checking out a memory of a previous nesting location.

MAY 30

An early morning paddle to mid-marsh. As usual, bird sounds dominate. Scores of tadpoles wiggle their torpedo-shaped bodies to escape over, under, and between pads of pondweed, many emitting muffled, burp-like sounds. Humorous entertainment.

The honkings of the resident Canada geese and the gulpings of the grebe broaden the marsh's collection of natural sounds. The Canadas are cloistered behind curtains of cattails, but I can overhear their gossiping. The repetitive calls of the pied-billed grebe remind me of a child banging the same note on a piano, again and again. It so happens I have my fly rod along, and it takes only a quarter of an hour to put breakfast in a bucket: yellow perch and one fat red-bellied sunfish will make fine fare.

I experience a moment of fisherman's delight, the type that leads to stories about the fish that got away: in a matter of only two or three seconds, with shocking velocity and strength, a fish rises, vacuums in my fly, and strips away my leader, fly line, and backing before I am able to react. I can only gasp at the suddenness of the moment as I sit, left with a lineless pole, staring at a swirl of water set in motion by the whatever-it-was fish—surely Nessie of Great Meadow.

I relay the incident to Barbara. She's heard it all before.

JUNE 19

Halfway across the Meadow's expanse there is an ancient buttonbush; picturesque, due to its jumble of curved and intertwined branches. A decade ago, before the last flooding by beavers, it was a strong, healthy specimen. Now it is seriously under stress from too much water around its roots. Its many dead branches are a favorite perch for herons and preening ducks; lower down in its trunky places, there are perfect dining niches for otters. Who could know? When otters bring fish out of the water, they often squeeze them between the slim trunks of such bushes, somehow making the slippery snacks more manageable.

What's that? Out in the Meadow in midafternoon I see, first with unaided eyes, an old stump—but, it seems the stump has grown. Time for binoculars.

A sight to behold! A large snapping turtle somehow managed to claw his way to the top of the stump and has draped himself a good foot above the water. It's not unusual for turtles to snooze, but this animal is totally asleep, completely relaxed. Four legs hang limp in full extension from under his carapace; broad head dangles loosely from a fully extended neck. He looks like a puppet that needs its strings pulled tight to get back into some recognizable shape. Ah, reptilian peace, available to top predators only.

The green frog chorus has more twanging and gulping members this year. The bullfrog contingent, with their bass voices, keeps moving from place to place in the marsh. Tonight, the groups seem to be at least one hundred yards south of last night's position. Their combined chorale is boisterous and fun to fall asleep to.

JUNE 22

Out on the deck, Barbara points and I look. There, flying by and unmistakable in the brilliant late afternoon sun, is a least bittern. Identification is easy, and a thrill; a threatened species seen here for the first time.

AUGUST 16

As I've not seen polar bears at Great Meadow lately, that lack was remedied by a trip to Svalbard, Norway. An unforgettable excursion, where we watched many polar bears engaged in devouring a dead whale.

Now I'm back to the Meadow to watch, not knowing how many interesting moments I missed during this summer's trip. The water level has stayed high due to past rains. The parade of rapidly growing goslings and ducklings continues, as they are taught the ropes of their lives to be. The members of the amphibian chorus have progressed from beginners to musicians of skilled note. Daytime sounds are dominated by the hums and squeaks of the pixie-frog crowd.

Like last year, seven female hummingbirds have made it an adventure to step out upon the open deck. Once they get plugged into the sugar drug at sunrise, or even before, any deck occupants can expect to be in for twelve to fourteen hours of general unruliness; avian entertainment at its best.

Ah, ha! Those free-for-all female hummers now have a male to contend with. Bold he certainly is, but when considered in human terms, foolhardy might better describe his determination to stake out a claim. He takes the challenge seriously. He flashes his red iridescent signal feathers repeatedly, shifting his head one way, then another before rising up on rattling wings to hold a stationary position in front of the ladies. He spreads his tail feathers to the fullest and flashes white tips, all the while sputtering his strongest vocalizations.

Usually, his audience spurns his conspicuous showmanship. This morning, though, all is not in vain. Appropriating one of the two feeders as his own, he feeds, retreats to a near-by spy perch, watches, and if one of the "girls" approaches his feeder, he dive-bombs her, scaring her away before she can take any sip of sweet.

The game is both time and energy consuming for the beleaguered "man of the mob." His defense can't last. By the third day, the girls win. As a result of constant, mettlesome activity, they force the over-matched male out of their realm. The deck area has now returned to its previous matriarchal wildness.

AUGUST 17

A great egret dressed the marsh today.

As I lie awake with the uncomfortable aches and pains of a summer cold, the night at least brings gifts. Three owls—barred, great horned, and long-eared—send messages forth and back, the latter voicing his soft, smooth, barely audible hoots somewhere just outside my windows.

AUGUST 22

The aura of the marsh is encompassing; it's a perfect evening. On stage here in late summer is a much-loved spectacle: cedar waxwings pursue insects, their lithe bodies trembling in the sunlight, lending their own definition to "grace on the wing."

SEPTEMBER 22

Summer's end.

Great Meadow is in its end-of-the-season repose. The water level has dropped about two feet and flats of pondweed have turned to shades of brown, adding soft, autumnal hues to the transitioning view.

SEPTEMBER 29

Midmorning, I watch as an osprey shows me yet another variant of its dazzling life.

First, it dives for a fish. No contact, but instead of taking flight immediately after striking the water, the bird stays there and bathes, just as a duck would! Following that scrub-up, it leaves the water, circles upward, zeroes in on prey, and dives again—but misses. Again, it splashes water with its wings and dunks its head to toss moisture onto its back. The osprey repeats this several times before spreading its strong wings and lifting off across the marsh, properly bathed, but perhaps still hungry.

BE IT KNOWN:
Through these moments, days, times, seasons, years, I know the riches I am privy to here at Great Meadow; I know the value of every scene and every moment.

OCTOBER 5

Very dry. Great Meadow's under parts are showing. I see beavers regularly, but expect they will wait for fall's rains before tightening up the leaks in their dam.

A white-crowned sparrow appears at a bird feeder.

OCTOBER 10

Drought ends with nearly ten inches of rainfall. Great Meadow is spilling over. The wet world of the Meadow's occupants greatly expands.

NOVEMBER 7

This fall's foliage season is a bust. Not only were the maple trees' reds tardy, most barely had any color at all. The usual array of painted leaves simply never showed, and October's leaf peepers were not only disappointed, they were pelted with rain. (Since records have been kept, this October has been the wettest month recorded—ever!)

After October behaved badly, November to date has dropped its image of dull, dreary, and bleak, offering up days of sparkling glory. Avenues along roads, hills, and mountainsides smile with yellows, rich tans, deep browns, and crimson reds.

As is usual here at Great Meadow, the sunrises and sunsets require my attention. Today, a north wind is brisk and frisky. Invisible fingers unzip, unbutton, tug, pull, tap, massage, and flip every vulnerable leaf still clinging to the trees, doing their utmost to dislodge them all. This is nature's striptease act at its best. Following each gust of wind, blizzards of leaves fly riotously upward into the sky. They twist, zigzag, wheel, turn topsy-turvy, somersault, sail, dive, or flip end-over-end, each seemingly reluctant to give up the last moments of airy leaf life.

From my elevated windows I look into and up at all the action. Following each gust, the entire surface of Great Meadow applauds with an eruption of visual accolades in the form of white-capped riffles that sweep across its breadth. Some show!

NOVEMBER 12

The oaks on the far hill, rich in reds, glowing like the embers of a campfire, set the backdrop for late-day action on the Meadow. Today, the beaver scene is worth recording on paper.

Often, a beaver pair feeds together; sometimes, a third may join the couple. A record has been set today. I spot four animals clustered at the base of a clump of shrubbery, then along comes a fifth. My fascination with seeing that many beavers feeding together hardly sinks in before I notice a tussock sedge a few feet away, swaying forth and back. The count is now seven! This beaver picnic must be comprised of two families.

As I think about my good fortune, two otters crash the party. Chaos usually follows otters, but they show no interest in the mob of beavers and slide away. Equally disinterested, the beavers never miss a chew.

NOVEMBER 16

A large black bear has paid his first call. Attracted to the scent of the birds' sunflower seeds, I catch him standing on top of the deck's railing on his hind legs, shaking the tall, metal pole that holds two of my largest feeders. I walk out onto the screened porch next to the deck and speak softly to the raider. The bear slides off the railing and grumbles past me, ambling his way down the stairs.

NOVEMBER 26

Hearing an odd sound coming from the area of the deck, I snap on the outside light. A large, black shape turns quickly into the large, black bear seen ten days ago. Tonight, the bear has returned with greater purpose. I stand quietly on the porch and watch as he grabs the half-inch steel pole and bends it down to a right angle where he can dislodge one of the feeders. When the bear has shaken and rolled all the sunflower seeds out of one feeder, he moves on to the second. He manages to dislodge the top of that feeder so the seeds conveniently spill onto the deck. I watch. A black bear's tongue is a wonder—twice as long as you might expect and flexible beyond belief. He easily laps away my four or five pounds of spilled sunflower seeds in seconds.

I return to bed; nothing to be done right this minute. The bear has gotten his midnight snack.

NOVEMBER 27

It takes me some time to repair the feeder damage and to re-bend, sort of, the steel rod. Now, instead of being straight, it's more artistic, like the graceful letter "S." Love those bears.

DECEMBER 2

The beavers remain punctual. I spot several late every afternoon, beginning around three o'clock. I haven't seen the record number of seven again, but this day's count is close at six.

I watch carefully to see just what they're eating. My binoculars reveal each beaver diving down and returning to the surface with large clumps of sod that seems to be comprised of root balls held together with muck. They hold the clods (one looking to be over a foot in diameter) in their front paws and meticulously sort out the edibles by squeezing out as much of the water and soil as they can, then chewing and swallowing the nutrient-filled rootlets.

I've poked through similar soggy leavings of beavers and have had a hard time finding anything green in the clods. Obviously they, like porcupines, have discriminating tastes, with salads greatly preferred (when in season).

Tree Swallow "one visit of hundreds per day. See March 2007

2006

River otters snort.

Black-crested cormorants grumble.

JANUARY 4

Four inches of soft but wet snow across the surface of Great Meadow's ice. Every track in that white surface retains the fine outline of its maker's foot. Eastern coyote and fox tracks are always of interest, but finding bobcat tracks again, for the third consecutive year, turns the interest level up a couple of notches.

If my old memory is reliable, the route taken along the western edge of the Meadow is very similar to last year's route. It's a good feeling to know a wild feline has apparently established a territory and is being sustained by it.

JANUARY 7

Coming back on snowshoes from the northern boundary of my neighbor's land today, I swing down through a boulder field where many turkeys have scratched through the snow, seeking acorns. I then head farther downslope towards the marsh.

Birds! Birds in the winterberry bushes and in the buttonbush clumps. Birds scattered above in the trees. Many birds flying, and they are all American robins! Eighty, perhaps more, are flitting about, nervously and silently. It must be the horse farm that borders the north end of the wetland, plus the fruit and seeds available around Great Meadow's edges, that sustain this large, over-wintering flock. Two decades ago, one would never have seen robins in the winter.

I plod home across the marsh, but en route notice strange markings in the snow that beg for my attention. Sometime, a bit earlier in the morning, wild turkeys apparently flushed or were spooked from the opposite shore. They'd flown across the marsh, landing about fifty feet from the near-forested edge, right here on snow-covered ice. I was looking at the theatrical evidence of phantoms, and I burst into laughter!

The snow revealed where every foot and feather touched down during landing. Each bird negotiated his touchdown differently. The smoother the landing, the deeper the telltale grooves in the snow (made by the turkeys' outstretched wings and tail feathers). Those turkeys had experience landing in snow on ice. They practically stopped themselves in midair before their feet even touched.

Then there were the klutzy birds, the novices. Their landing marks revealed few deep tail or wing markings, but showed deep furrows where the turkeys' chests and bellies plowed into the snow. They obviously came in too fast, catching their feet in the drifts. That mistake pushed each one forward like a snowplow. No doubt more practice is needed. I spend a half an hour smiling over the stories I read in this snow.

FEBRUARY 5

Saw beavers today, but it concerns me that this is February. My sighting is indicative of this winter's "not right" weather—little ice on New Hampshire's lakes, with their deepest areas remaining unfrozen. Ice covers most of Great Meadow's expanse but high temperatures and many days of warm rain have opened up water around its edges, thus explaining the appearance of the beavers. Not a bad thing, because now they can travel out of their ice-blackened house and find fresh bark to eat. Normal February weather would keep them ice-bound in their lodges.

FEBRUARY 12

The winter doesn't feel right with these above-freezing temperatures and little snow. The only ice is thin, too fragile for my weight. Open water around the shore's edges has appeared at least four weeks early. Heavy rain on the tenth of the month liquefied what snow there was on Great Meadow's surface. That was followed by below-freezing temperatures on the eleventh, and a welcome surprise this morning—a skateable surface.

By midmorning, I am on my blades, hearing old muscles creak, feeling them complain, but the exhilaration of skimming over the ice, faster, faster, leaves the complainers behind. Friends appear across the Meadow by the beaver house. In my haste to greet them, a blade catches in a crack of ice, tumbling me in an ignominious heap at their feet. I hide the fact that my shoulder is yelping with pain.

Snow is forecast in late day, so my skates go back onto the shelf in the garage, back to their job of collecting dust.

FEBRUARY 20

Last week's snow melted due to the persistence of unseasonable warmth. I put ice cleats on my hiking boots and walk across the Meadow, staying conscious of thin ice.

The sun beckons and a meadow-edge field invites me to spend a while, even to nap. A bed of moss at the base of a large tree is mattress enough, and the sky offers a soothing winter blue. With a wrap of winter silence around me, there in my own cocoon of crisp, clear February air, I drift off.

MARCH 9

A red-winged blackbird arrives, but this early? After a winter of discontent, when Great Meadow's climate-related abnormalities have been a constant, this morning's unseasonably early call of the red-wing sounds a wake-up alarm throughout the slumbering wetland. The subliminal spirit of the Meadow senses the calls of spring in a variety of ways.

MARCH 15

Common grackles return.

After a week of winter's coldest weather, the ice cover has finally thickened, but too late. Its life will be short. Already, throughout the Meadow, the larger clusters of aquatic bushes show open water around their trunks. The darkness on each stem indicates that the sun's heat has been absorbed over several days, acting as a buffer to ice.

Those gaps somehow transmit messages to the beavers in their dark lodge. First thing in the morning there comes a beaver, lumbering up and out of one of those openings, having traveled over three hundred feet under ice to get there, like the otters.

So many mysteries surround those superb animals. How, for example, do they sense the long route under ice, from the lodge to that small patch of open water? This morning's beaver had to swim under the ice, negotiating a labyrinth of twisty channels, from the underwater exit of his lodge all the way to this one opening in the ice. What combination of senses acted as its guides? Anyhow, he finds sustenance in gobs of muck, as the beavers did at the beginning of the winter's cold. He eats for about an hour, returns to the opening, and, renegotiates those three hundred feet back to his home base.

MARCH 25

With ice openings widening, yesterday brought hooded and common mergansers back to Great Meadow. With the common mergansers, females returned first, a pattern noticed in previous years.

MARCH 28

Again from the tropics, those grand swirlers-above-waters, the tree swallows, are back.

Come afternoon, Great Meadow provides yet another spring gift. When I walk into our great room, a shadow sweeps over the oak trees. Something big. I dash to the largest window and see that the shadow-caster is joined by a mate; perhaps a family member.

Two bald eagles perform tight circles and swoops in what I hope might be preliminaries to an aerial mating display. One's heart stops beating when witnessing those body-twisting, upside-down, talon-grasping moves, especially when they are executed only yards away at eye level.

The eagles separate. One bird flies off toward the north end of Great Meadow. The other lands on the Perch Tree. After ten minutes, the nearer eagle apparently spots something in the water on the far side of the marsh. A few strong wing strokes, then a long glide, carry the bird to a place just over a stretch of ice that has yet to give up winter. The eagle stalls in midair, then drops into a long, narrow slot of open water.

The instant talons sink into prey, it seems as though the eagle has flown into an invisible net that's hidden just below the water. Instead of the bird snatching the prey and flying off, the prey seems to have reached right out of the water and grabbed the eagle. The fish struggles, shaking the eagle, but just for a moment. With eight talons embedded in fish muscle, the eagle quickly immobilizes its prey.

But the fish is too heavy. Every attempt to get back up onto the ice causes the thin edges to give way. The eagle's great wings spank the surface again and again, in an attempt to row forward and gain the support of thicker ice.

Every eight or ten flaps move the eagle about five feet through the water. After each of these energy-zapping efforts, the bird stays motionless, sunk deep in the frigid marsh, wings extended to stay afloat. As the eagle rests in that icy water, I wonder about the insulating qualities of feathers and how long they can stem hypothermia. I am uneasy.

Six sessions of wing-whapping are necessary before the eagle finally reaches ice that's thick enough to hold weight. It pulls itself up onto the ice with one foot, still clutching the fish in the other one. That's when I first see the prize—a white sucker, at least fifteen inches long, weighing probably three pounds or more!

Once on stable ice, and as if there was no ordeal getting there, the eagle immediately begins to rip into the fish's choice head parts. Although I had been most concerned over the great bird during the struggle to gain a solid footing, I now wonder if the predicament had ever been serious. It may be that bald eagles can quickly judge whether or not its catch is too heavy to lift, and if some space, land, log, or ice is nearby, it is only a matter of dragging the prey there. If the prey proves too large, or there is no place to move it to, I expect the eagle would simply release the catch and return to the hunt.

With the hard-won prize roughly half consumed, the other, forgotten eagle sails to a bouncy landing close by on solid ice. "Not sharing!" screeches the prey's owner, head turned, hackles raised, mouth agape. The standoff continues as the would-be meal-crasher first walks around on the ice, then takes to the air again, circling and swooping in low over its competitor in a return to the dinner table. This induces a second shriek of defiance from the fishing champion—the shriek followed by another bite of fish.

A second fly-around follows, the interloper buzzing even closer to the head of the fish owner, who this time takes the threat a bit more seriously and grabs a large chunk of flesh before flying away to a perch on the far side of the Meadow. The remains of the hard-won fish provide the uninvited diner with a relatively easily won meal.

APRIL 9

Wood ducks have returned to Great Meadow to join the throngs of spring arrivals. Now, I wait for ospreys.

APRIL 17

A long wait, but yesterday's air over Great Meadow supported a single osprey. On first sighting, this bird cut, turned erratically, dipped, glided sharply upward into some accommodating movement of air, quick-stroked, and slid sideways over to the far shore, dazzling my eyes anew with perfect aerial acrobatics. Few spring tonics can equal that; not even a martini.

At some point in time, New Hampshire became the birthplace of this osprey (of course the "when" is not known). While its winter was spent someplace in South America, the long route to Great Meadow has become encoded in its body. Coastlines, Earth's magnetism, stars, scents, colors of certain waters, densities of air, position of the sun, some sense of longitude —who knows the wonders of inheritance that guide such endowments.

MAY 9

When beavers entered the Meadow and constructed their first dam thousands of years ago, the brook flowing into the nearby pond slowed, and any sediment carried by that water settled to the bottom. The habitat evolved from a deep kettle pond to a shallow meadow.

Now, a bit of more recent Great Meadow history: In 1980, a farmer whose land abutted the Meadow became incensed when he found that beavers had been going onto the edges of his pasture and taking down saplings. In a hardened response, he first had the beavers eliminated. Next, he drove his land excavator down to the beaver dam and demolished it completely. One hundred acres of dammed-up water rushed through the raw opening.

Nature leaves no suddenly opened space vacant for very long, especially one with rich, hydric soils. Seeds from wetland-loving plants took advantage of the opportunity to fill the spaces and grow.

Grasses, forbs, cattails, reeds, bulrushes, arrowheads, pickerel weeds—all found space to thrive without competition. Those plants were not alone, as hardier, wetland shrubs, like leatherleaf, sweetgale, and buttonbush, also flourished in newly found niches.

Now, two decades later (the objecting farmer having since passed away), the toughest of those plants, the shrubs, struggle to survive the re-flooding work of beavers. As tough as those species are, few lived beyond another decade. Only the crowns of buttonbush and leatherleaf find sunlight above the water's surface where the water is deepest.

Every new family of beavers that has reconnoitered this meadow and

Beaver sculpturing

Critical questions to be considered for a successful,
sustainable beaver flowage:
What is the quantity and availability of the supply food?
Will the depth of water, once a dam is built, be sufficient to give the
lodge the safety and survivability necessary?
What of the quality and quantity of the incoming water supply?

brook system over millennia has arrived at the same conclusion on where best to construct a family-supporting dam. When a new family of beavers migrates up the outlet stream, they "read" the lay of the land and its hydrology. Today, the average depth of a flooded Great Meadow is less than two feet. After many tomorrows, a swamp will replace today's vegetation and trees will grow where herons now fish. The brook will continue to flow, playing other roles in its relationships with its surroundings.

MAY 10

The herons are here and I have heard a pied-billed grebe call. Spring peepers are sounding in numbers most satisfying. Pickerel frogs orchestrate their own roles in the chorus. Missing and missed is the music of the resident robin. Ah, but the northern oriole twiddles and whistles.

MAY 14

Yikes, I'm seventy-seven! A matter of no consequence to the Meadow, although the continued deluge must be.

Heavy rains over the past three days have triggered the gears of living plants and animals to shift, while the local streams and river exercise their waterpower. Towns have been isolated. Great Meadow's brook parades its muscle, and changes the Meadow into a lake.

The tree swallow houses, elevated three feet above the Meadow's natural level, now have water threatening their basements! A hooded merganser female is known to be setting in one of the duck boxes, with a wood duck mother in another. Both are flooded. Yet another example of well-intended, but obviously misplaced human tampering with wildlife.

Furthermore, any ground-nesting waterfowl that have built shelters on Great Meadow prior to this rain will be lost. Many concerns arise about the local wildlife during this season of birth, and of course my mind shifts to the water pressure being exerted upon the beavers' pile of sticks that hold back the Meadow's one hundred-acre flowage.

MAY 22

The ingenuity of beavers stands out again during a visit to their dam. Hundreds of sticks of varying lengths and widths, globs of sod, cobbles, and rocks, each woven, packed, jammed in, placed, pushed—every piece of chosen material intended to hold back their pond defies the deluge.

As dams go, the structure here at Great Meadow is of the smaller variety, but size in beaver dams has little to do with how much water they are capable of impounding. Dams must provide basic needs, and the beavers' structures may meet those requirements in any given location, even where the actual flowage might be only an acre in size. With each dam, with each flowage, the beavers' sole purpose is to provide for their families. They know nothing of the supported community created with every dam they build.

Great Meadow's beavers have formed a dam roughly thirty-five-feet long, but its genius lies in its shape: it curves upstream against the push of water. Across the narrow opening at the outlet of the marsh, the dam is engineered to withstand water pressure as mighty as was exerted during last week's flooding. With its ends solidly laid against firm earth and rocks, the dam's bundle of sticks withstood this May's hardest punch.

MAY 30

Symphonic nights surpass the intensity of amphibian sounds heard from Great Meadow's orchestra since our coming in 2000. Tree frogs and spring peepers dominate the chorus, their rich notes combining to supply the coming night with a crescendo of sounds that twirl in the darkness like spellbinding music. It's as if the night air is liquid, a stream if you will, filling listeners with the sound of voices newly emerged from the depths of the marsh.

The rapturous sounds, as full as they are, represent but a miniscule portion of Great Meadow's depth of life and breadth of diversity. Like wetlands everywhere, Great Meadow's flora and fauna are climate-dependent, with some years productive for progeny, some bad. Over thousands of years, most have adjusted to some level of dependable average.

The notes issued into the night symbolize hope that the future will be guaranteed for the singers, that the roles played by the various species of frogs and toads will be repeated year after year, and that their contributions will keep the marsh sustainable, both for themselves and for others that will eat them.

The year has turned, their time to mate has come—and it came the night before and the night before that—but now, tonight, may be a final chance. Tomorrow will come and pass quickly, progeny will take their place. As with all species, including our own, life's calendar always has too few pages.

Species that cannot adjust become extinct (mankind now plays a fast-increasing role, mostly indirectly). But what happens when nature's averages are either no longer dependable, or those formulas shift enough to make adaptation too difficult for some species? My observations indicate that some species at Great Meadow are struggling to adjust to climatic conditions that are obviously changing and resulting in altered arrival times of migrants passing through both ends of the seasons.

I also notice changes in species native to where I have lived for these many years, several having vanished from the my area completely. Species new to the locale keep appearing. I feel the losses of those that were always here far more deeply than I celebrate the gains of new arrivals. The most disturbing shift of all: with few exceptions, it is the loss of quantity that troubles me most.

MAY 31

Wood duck mother with ten ducklings! Due to the recent flooding, I'd assumed most nests of geese and ducks would be lost. I am more than pleased to see this raft of tiny paddlers. Of the ducks that nest in this geographic area, it would be wood ducks that are the least affected by high water, as they nest, if possible, in cavities they find in aging trees.

JUNE 3

A second surprise: A successful fledging of two Canada geese pairs, together numbering sixteen! The two families move as though they are constrained to follow a single track. Their strictly linear formation is a sight to behold.

This particular train of geese consists of one female in the lead, followed by a perfectly spaced line of five goslings. A male comes next in the line, and directly behind him are gosling numbers six and seven of that brood. With no spacing behind gosling number seven of the lead family comes the female of the next crew, with four goslings following her. Then comes her mate, with a single fluff of yellow bringing up the rear of the train; a tiny caboose. Those sixteen geese could not have moved in a more precisely formed animation, the sequence of two families keeping their snaking formation in perfect order until a bank of trees hides them from my sight.

In spite of my delight, seeing what appear to be two families of Canadas paddling along amicably isn't necessarily the norm. Compatibility among and between goose families is not guaranteed. Prior to, during, and following nesting, families run the gamut from relative peace to outright conflict. Couples with young goslings are usually very protective of their charges, and they have serious reasons to be: jealous ex-spouses, independent males, or females are apt to do severe harm to unprotected nestlings, even going as far as to destroy eggs in nests left unguarded only momentarily.

It could be that these adults are babysitters and not the parents. It is not unusual for a Canada parent to give temporary custody to a surrogate. There are no known explanations for such behaviors. Whatever the depth of the disciplined coziness of the two families seen parading across Great

Meadow, it's good to think that peace can come with good relationships.

JUNE 4

More heavy rains and flooding have returned, and Great Meadow is inundated once again. Water has risen to cover the duck box entrances, but the time for harm has already passed.

JUNE 10

On all but one of the last six days, clouds have backed in from the southeast and flushed their loads of water over southeastern New Hampshire. This deluge, unusual for this time of year, is seriously affecting the state's economics and agricultural pursuits. My few pots of flowers are little more than tiny ponds holding drowning occupants. My planted seeds are mush.

Somehow, a pair of robins has managed to build their nest in one of my next-to-the-house peach trees. This morning, the egg-layer sits stolidly on her eggs as rain beats down, challenging her commitment to bring life to embryos come "Hell or high water." It seems both have arrived.

JUNE 18

It's 99°F by late this afternoon. The climate is in yo-yo mode.

The orange, black, and white of Lord Baltimore's bird distracts my attention to Great Meadow as he flies forth and back. For most of the past month, "my" northern orioles have been busy restocking their bodies following their journey from the American tropics. My mornings are filled by the male's music, and my late afternoon sips out on the deck pulse with spring's life, enriching me with song and sight.

The choice of the orioles' nest site is closer to the house than last year, but in the same red oak tree. True to their inheritance, nest sites for the orioles' unique, bangle-like construction are high off the ground, apt to be located at the far end of one of a tree's longer branches. Genes partially dictate the blueprint for the free-swinging basket of beak-woven sewing.

These birds frame their nests by hanging plant fibers, strips of grapevine bark, and coarse hair over branches, leaving a type of hammock swinging in place. The female clings to those suspended materials like a trapeze performer, somehow crafting an outside framework for the sack. When her weaving is finished, she pulls together the longer fibers to cinch the lower part of the nest taut.

Northern orioles decorate their nests from the inside. Today, the female brings materials that look white; perhaps dandelion seed fluff. For the next two days, she will be plainly visible through the sides of the nest. By the third day, she will have added enough binding fibers to close that window so that only when the nest is wiggling will I know she is still at work.

The nest is hidden from possible predators from above by a canopy of leaves that overtop the opening like an awning. The leaves growing along the entrance to the nest have not been disturbed. The opening is small and requires the adults to enter headfirst; feeding is always done in that upside-down position. Construction of the egg cradle lasts about a week.

After the chicks hatch, their world becomes a grassy sling protected from rain and rocked by gentle breezes; how lovely.

JULY 30

The Meadow draws my eyes this morning. Twenty, eighty, perhaps more times each day, my mind draws sustenance from Great Meadow: light, shadows, tapestries of weather, patterns of plants, and incidences of life. One duck or forty, an osprey stationary in the sky or in a spectacular dive, a skein of Canada geese or a solitary barking goose, a log topped with twenty sunning turtles, a dragonfly balanced on a tippet of grass—every sight is a gift.

Repeated raspy calls from a great blue heron, who is particularly vocal this morning, accompany the waffles Barbara serves on the porch. Buzzing hummingbirds vying for sips at their feeders fill any silences in our conversation. Nice how natural sounds seldom disturb.

AUGUST 2

As the crust of a wild blueberry pie takes shape under my rolling pin, a quick glance at the Meadow brings my baking to a halt. For the umpteenth time, something catches my eye on the treasured Perch Tree.

Due to the morning's heavy, humid air, a ground fog slumps over the surface of the Meadow, but leaves the entire top portion of the tree exposed. There, through the air's bluish haze, are three grand birds: Two great egrets and one great blue heron. Each bird is busy preening, and each feather lifted by a beak reflects sunlight for a brief twinkle. The whites of the egrets, the blues and grays of the herons, all suspended in airy pastels, could have been easily dismissed as an apparition. For those once-in-a-life-

Water lilies on Great Meadow

time moments, a piecrust can wait.

AUGUST 3

Something brings the great egrets to Great Meadow about this time every summer. They fly inland from the coast following some internal beckoning. Again, my belief in individuality is reinforced. These birds flew here due to a decision each made, or due to the action of another egret. I can only guess that some aquatic life becomes available at this particular meadow at this certain time of year. How that opportunity is communicated from Great Meadow to these lovely, white birds is an intriguing unknown.

I watch them carefully. They feed along the mucky edges where the detritus of the marsh has built up sufficiently to create long windrows of accumulated debris. The birds purposefully walk those lines. As their long, snaky toes sink into the dark softness, their beaks stay poised, ready to snap up some tasty life hiding in the mud.

AUGUST 7

Five moving shapes of virgin white pierce the mist of this morning's steady rain; great egrets arranged at various levels on the Perch Tree. They look like notes on a staff of music, a chord really, as the birds have chosen not only separate limbs, but they all navigated to the left side of the tree's main trunk and stand at slightly different levels. The notes being played: C E F# A C!

AUGUST 29

The encompassing arms of Great Meadow extend outward from its wet edges into and onto its 360° of varied forest. My steps have touched the edges of that full circle, perhaps one hundred times. This year, I've concentrated on the edge I cleared between my home and the Meadow.

Keeping the active botany of forests away from cleared spaces demands work. Not so hard if the cleared space is a lawn, where periodic whirling blades can rip off any green shoot that answers to the sun. The opening I've created between the Meadow and my home has been purposefully planted mostly with the needs of birds in mind—many types of berry bushes, three kinds of grapes, a couple species of dogwoods, winterberry, red elderberry, and others.

My plantings have paid off, as the space has become a much-used habitat for nesting songbirds, the ground space attracting a variety of small mammals and reptiles, and there is tender browse for larger animals like deer. Maintenance is a challenge, as keeping down a forest of wannabe trees on a steep, boulder-covered slope is demanding for legs that are ever more reluctant to do what they once did so well.

While I was cutting back daylilies yesterday on the marsh side of the house, there on the downside of the second terrace, a long, dark shape of a sizeable northern water snake materialized. She was quite a distance and a long slither from her preferred watery world. Her extraordinary array of senses must have somehow informed her that lunch—maybe a chipmunk or a mouse, a fine meal for this Meadow dweller—might be available on the rock terrace high above. Good to make her acquaintance. I'm glad she is here.

Hummingbirds are still here too, scamps as ever!

OCTOBER 2

A narrow slit opens in the bank of clouds settling down for the night over the western trees. The sun pierces that opening, and October's intense light illuminates Great Meadow, turning its black water to shades of green and blue, highlighting the wetland's parade of autumn trees in its reflections.

As my mind is dizzied by the sunset colors, Great Meadow's stage presents a full production. Flying low over the sunset-struck water, an osprey pumps its way toward the Perch Tree. Dangling from its left talons is supper. So bright is the sunlight, my eyes play tricks with the hawk's identity. The bird's whites seem to have spread to places they should not have, and its blacks sparkle to the extent that they look liquid.

The routine so expected from an osprey on its perch with food is not followed. The bird fidgets, holding its head, with those scarlet-red eyes, high and steady—oh! for good reason. A bald eagle comes down-meadow from the west, riding a beam of sun like a projectile, with clearly one objective: the osprey's meal. *Seize it!*

As the distance to the target closes, the eagle suddenly, surprisingly, makes a sharp course correction. Three powerful reverse wing strokes almost stop the predator's forward motion. A turn of its tail feathers banks the bird into the required angle of attack. Cupping broad wings to create further drag, it drops into the water, and yanks out a fish.

Now, one-upmanship of predator against predator begins. The outcome is never in question; size matters in this confrontation. Accepting that its supper must be eaten elsewhere, the osprey vacates its position on the Perch Tree just before the eagle arrives on the same branch with its own meal, not a stolen one. A chance opportunity at the last second changed everything.

The clouds in the west close the curtain and only dim light remains, but during those couple of spotlighted minutes, I understand that what was grand drama for me was a normal part of life for the actors.

OCTOBER 15

Every year, predictions pertaining to leaf color flood the local news outlets. As is usual when the so-called experts pronounce their weather-related guesses ad infinitum, they are mostly wrong. This year, the prognosticators are one hundred percent off base. "The spring was far too wet." "Plants and trees too stressed by rain and cool temperatures." "Summer was much too cold." And on and on.

I will match this year's fall brilliance throughout New Hampshire, and certainly here at Great Meadow, with any other season in memory.

2007

Tree swallows gurgle. Green frogs gulp.

JANUARY 12

December's unusually warm weather topped all-time records. Not only was there no snow, thermometers squeezed well into the seventies and even low eighties through Christmas and into the first days of this year. Only last night did the jet stream begin to shift, forced southward by a high-pressure ridge from the Arctic.

Beavers and otters have been on the Meadow regularly, as if October conditions still held sway. Yesterday, eight common mergansers, all female, and seven Canada geese behaved like it was still fall.

Anomalies with weather can be either positive or negative. Such strangeness has always been with us, but when atmospheric conditions begin to cause abnormal weather (too hot, too cold, too dry, too wet) for too long, at the wrong times, my interest changes to concern and my concern shifts to wildlife.

JANUARY 20

Finally, a chill in the air makes me hopeful that winter has arrived. Last evening, a tardy messenger of cold touched Great Meadow and stilled its surface waters with a thin sheet of ice.

JANUARY 28

Too-cold-too-quickly has stayed all week. With a well-below-zero wind chill, the Meadow's surface water cooled to freezing in a few days, instead of over a period of several weeks. No harm to the living water, but how many of the wetland's multitudes were able to adjust to such a quick freeze?

I feel the ice on Great Meadow calling. Ordinarily a snow skim might make smooth ice a challenging surface for walking, but apparently the snow that fell late yesterday was quite moist. When it floated down onto very cold ice, flakes were instantly glued into place. While there are isolated slick spots to avoid, I can walk along at a normal pace.

Ah, finally, a stimulating winter morning. The marsh has new stories to tell. A fisher (a large male) and a red fox checked for mice in rock crevasses and root openings along the shore. With fine tracking snow, the animals left their activities clearly imprinted. To interpret the "language of feet" with some degree of accuracy, you must first be where the feet were. Second, you must learn how to think like the animal that left the story for you to decipher. (A respected friend, Susan Morse, founder of Keeping Track, Inc., in Vermont, has been my mentor in reading wild footprints.)

Today, the story is clear. A small pack of eastern coyotes came across a young deer that had made the unfortunate choice to bed down in thickets close to the Meadow's shore. When the deer panicked, the coyotes quickly forced it out onto the ice. Once its hard hooves came into contact with the slick surface, the deer was doomed. As usual, the coyotes dealt efficiently with their prey; only the rumen, clumps of hair, and a bit of skull remain.

I come upon muskrat lodges at the north end of the Meadow. Good news, as they have been scarce here for several years. I am always pleased when I happen upon one of these foot-long, industrious furbearers. Muskrats prefer to snooze during daylight hours, but chores must be tackled come dusk. Early spring means it's time to fatten up on tender grassroots; during summer's indolence, they build discreet feeding platforms along the shoreline out of mud and middens; and fall brings demands to prepare accommodations for the season of cold.

Today, I count nine lodges, six of which are located along the western shore. The size of the structures is impressive: two are six feet wide and a bit over four feet high, roughly twice the normal size. No answers to the question "Why?"

FEBRUARY 3

Having seen hundreds of muskrat houses, I tap into memory and note that two modifications in engineering seem to have been happening over the years. First, many lodges have become larger in total mass; and second, instead of building their structures in fairly open marshes among emergent plants in shallow water, some lodges today are being built to encompass the slender trunks of surfacing woody bushes, such as winterberry. The lodges so enclosed might have an advantage, I suggest with caution: the stout stems of those shrubs prevent dogs and eastern coyotes from digging into the structures.

FEBRUARY 8

Beaver update: One family has built a new lodge along the far eastern shore, about one hundred yards east of the Meadow's primary dam. I expect the old lodge in the south cove has been abandoned. The second family, calling the northern half of the Meadow its territory, has also abandoned last year's quarters and a new, larger lodge now squats in the top third of the meadow. Did any communication to rebuild pass between the families?

FEBRUARY 11

A frigid week ends with a warm sighting: a red-bellied woodpecker, a first here at my feeders. He arrives sporting his brilliant zebra-like back, his striking pinkish head, and his faintly rouge belly feathers. The male spends several minutes checking for edibles in a dead oak stub, one partly demolished yesterday by a pileated woodpecker.

MARCH 2

Red-wing blackbird! Too early. . . . Too early.

MARCH 22

Ice still occupies Great Meadow. Last Friday's fourteen inches of snow followed by rain sealed up the openings in the ice that the otters had been using.

I spot a great blue heron, and the red-bellied woodpecker appears almost daily. That's good. A great blue heron stands on the marsh in snow. That's not good.

Every morning, I spread chunky peanut butter across slices of stale bread and sail them, one at a time, off the deck onto the steep slope below. Red squirrels, black-eyed juncos, and blue jays race to those offerings, the crowd arriving almost before the food hits the ground. Not unusual for four or five blue jays to squabble over a single slice.

During today's early morning feedings of the many furred and feathered breakfast comers, another treat greets my eyes. As several jays attack a slice, the red-bellied woodpecker drops right into the middle of their party and disperses the crowd. A bit later, he rousts a second gathering. This bird has chutzpah; let's hope his maleness will also be attractive to a lady.

MARCH 23

Make-A-Difference-Donor award received in Manchester today from Council on Fund Raising (CONFR) for donating time and resources. I spoke at the ceremony.

MARCH 24

Two spring gifts today.

In the morning, I walk out on questionable ice to clean the tree swallow nests that stand on posts in the marsh. During breakfast, I had remarked to Barbara that we should be seeing one of those birds any day now. As I step out onto the ice, there, perched on the nest box pole, is a tree swallow, sitting unafraid, offering soft, gurgling phrases until I am almost within touching distance.

I clean the box, strengthen it. To accommodate the crowds that always gather around, I attach a long, dead branch to the house. As I turn to walk away, the swallow returns, now perched atop the offered branch. The bird seems to bubble all over with energetic twitters.

As I clean and replace several other nest boxes, the ice, in its advanced age, refuses on two occasions to bear the burden of my weight. Not pleasant. It's not that the water is deep; it's the plunging into marsh muck up to my knees that elicits a few words seldom used in social circles.

Late in the afternoon as Barbara skunks me at cribbage, I spot an immature bald eagle high in a marsh-side oak. Its plumage radiates richness due to an abundance of white feathers accentuating the browns on this late-adolescence bird. The undersides of its tail feathers already show white. Its head and neck sport far more white than brown, and its entire chest and even most of its back wear checkered colors.

After a half hour, the eagle drops from its high branch, sets its grand wings, and without a single stroke, sails far out over the meadow, landing on the ice next to a buttonbush clump. The eagle remains perched there even as darkness falls.

MARCH 29

An early spring riot of frigid air tested every tree's integrity yesterday, and continued to raise havoc throughout the night. Morning's first light reveals that Great Meadow's wonderful Perch Tree failed the test.

It's now horizontal, crumbled into a broken trunk and shattered branches, branches that have, since our coming here in 2000, held scores of eagles and ospreys, herons and wood ducks, egrets and crows, kingfishers and gulls. The tree that I glance at every time I set my eyes upon the Meadow is gone, a major loss to this watcher, and to every viewer of this extraordinary place.

MARCH 30

Phoebe is back! An osprey, a pair of wood ducks, and a bald eagle curve around the marsh regularly, but without the Perch Tree I only see them in flight. That fallen tree had been my eyes' first destination as I scanned for Great Meadow's flying hunters. Its loss is significant.

MARCH 31

Two actors against a background of sky require my attention: A pair of bald eagles, one immature, one adult, are practicing aerial sparring. Wide, banking turns in slowly tightening circles bring the birds near collision every time they pass one another. I expect I am privy to the teaching of preliminary lessons that, one day, will become truly mercurial aerial maneuvers executed by mating bald eagles.

Eagle watchers who can anticipate the next exciting maneuver never shift their eyes away from the action. One, two tight turns, a swoop down from above, another upward swoop from below, and suddenly one of the animal kingdom's most dramatic moments—the upside-down flip. The mature male somersaults underneath the younger bird. Talons lock onto talons, immediately neutralizing both birds' airworthiness. With talons engaged, the birds drop into a downward spiraling tumble, each holding their great wings in a half tuck, plummeting ever closer to the ice covered marsh.

At what seems beyond the last split second, talons release and broad wings grab air as both birds barely clear the ice. With certain death defied, the acrobatics end. Then the adolescent seems to flee, with the adult in hot pursuit.

No way to know if the scene was fun or serious, but no matter—it was spectacular!

APRIL 1

Common mergansers; three males, two females.

APRIL 2

All day, the hooded mergansers and wood ducks have been milling alongside each other around the three nesting boxes I installed back in 2002. The squabbling over those nesting sites does not change, nor do I cease to be enthralled by the competition. While watching the winged skirmishing, I wonder if there might be some carry-over memory of the loss of all three nests last year. I expect not, but to prevent repeating the flood debacle, I've erected three new nesting boxes and placed them high on tree trunks at varying distances from the marsh.

APRIL 8

My calendar reads Easter, although this year, the hoped-for April showers call for woolen hats, not Easter bonnets. Snow is on the way.

The marsh has mostly refrozen, leaving only narrow reaches of slush-free water. Quite a scene, as Great Meadow's early migrants are restricted to those limited spaces. Geese and ducks form in floating crowds of what appears to be total compatibility.

In midmorning, an eagle looks for, and finds, frozen carcasses of fish stuck in the ice. Naturally, as it rips away meaty chunks with its beak, a trio of crows cannot resist joining. Each crow caws and strolls, always behind the feeding eagle. The eagle ignores their truculence, finishes the cold snack, and leaves.

Yesterday, after finishing up domestic chores and letter-writing duties, I felt that a walk was in order. I trudged through crusty snow to where our property to the north comes to the shore of the meadow. An osprey was circling and crying as it flew along the Meadow's edge. Surprise! It chose a tree limb just over my head for a perch. Our eyes met in what seemed a meaningful way, but the exchange ended there.

Having just read David Gessner's *Soaring With Fidel,* it occurs to me that this Great Meadow bird could well have just returned from northern South America via Cuba. The mere thought of such an animal, perhaps this animal, having flown all those miles south, plus all those return miles north along coastlines and over expanses of the Atlantic Ocean, gives credence to the sense of appreciation and respect I feel for its remarkable life and capabilities.

A gust of wind invites the osprey to open its wings, and the "fish hawk" accepts.

A bit farther west and uphill along my neighbor's boundary, my old heart requests a rest. In that interlude, a pileated woodpecker swoops close by, landing on the trunk of a dead American beech tree. A female, and from the size of the cavity in the tree, I can see she has been working this site for some time. I watch her persistent labors, hammering, hammering, with wood chips flying. The power exerted by each strike can be appreciated simply by holding one of the chipped out pieces in your hand. This species creates a new nesting site every year, the old ones becoming of great benefit to owls and tree-nesting mammals.

While watching the woodpecker, I catch a movement to my left: deer walking, stopping, sniffing, coming directly towards me. Two are close together, with three more slightly to the right. At about fifty feet, the second deer in line sees a forest shape he's unfamiliar with. (I'm downwind, so the caution had to have come from shape, not scent.) From that second deer's reaction, a message of caution is somehow transmitted to the others.

No sound is provided for my ears, yet all movement stops. One deer actually remains in mid-step, one hind foot suspended. Five minutes pass. The woodpecker continues her hammering. Man and deer remain motionless. More minutes tick away.

It's not often deer lose patience before man, but eventually, the closest deer raises his front leg and stamps his hoof sharply into the ground. Uncertainty is giving caution a nudge. I break the silence by squeaking like a chipmunk. That confirms to the deer that his wariness best be acted upon. No panic, but they exit with flawless grace into a thicket of trees, their raised, bobbing white tails the last image I see.

My attention again turns to the den-building woodpecker. The cavity is about forty feet above the ground in a large tree. The shape of the cavity is oval, and that's of special interest to me. Pileateds are seekers of black carpenter ants, a primary food. The ants carve out tube-shaped chambers to maintain and grow their colonies in the central part of tree trunks, often in masses of several hundred.

With the high probability that a nesting hole is being made, I'll be back.

If I time that visit correctly, I hope to confirm something I'd witnessed about these highly fascinating birds several years ago.

APRIL 20

Great Meadow's stage features new acts every day (really every minute), especially at this time of year. I've witnessed two performances within the last half-hour.

The first: thirty to forty common mergansers, all creating their usual turmoil, although it's a particular male and female couple that draws my attention. The male comes up with a fish; the female wants the fish; the male flaps and quick-paddles across the water's surface with the fish; the female follows in pursuit. Together they twist, turn, and spin, but the prize holder never gives up his catch.

Strategy change: The male dives, the female dives—but surprise: The female comes up with the fish! Now, the chaser is the one being chased. Flapping, speed-paddling, twisting, and turning across the water, the female holds the prize high in her beak, then dives. The male follows.

Ah, now the male comes up with the fish! The chase resumes until he decides to halt the race by tossing his head upward and swallowing the booty.

Shortly afterward, six ospreys spot fish close to the surface. The birds dive almost in unison, thrilling me with their mastery of control. I wonder, can my appreciation reach a limit?

APRIL 24

Following an afternoon of garden work, the deck looks inviting so I answer its call with an iced coffee, a book, and of course, binoculars. I never get to open the book.

In two marsh-side trees, collections of various blackbirds are in full concert. The mixture of red-wings and common grackles burble forth a cantata of liquid notes so sweet I dare not let my listening stray for a moment. A dozen birds in one tree, and thirty or more in the one next to it. A breeze sets a steady pace from the northwest, so every bird faces in that direction. The *a capella* tune includes rests in its measures. Each male bird seems to be in full complementary voice, red-wing with grackle, grackle with red-wing. On occasion a single bird flits from one perch to another,

Pileated woodpecker's delight: black ants for lunch!

but mostly each singer stays in place in the choir loft. The chorus is mellow and charming.

Most years, these two species of blackbird perform this concert in tandem a few days before the females arrive. It's like a boys' good-time barbershop concert before the serious business of protecting nesting sites and raising fledglings gets underway. When family duty begins, singing time in the treetops is over.

The wonder of these ear-pleasing minutes of music ends when some signal sends all birds but three into the air. As if their flight is controlled by some invisible source, that flock surges out over the Meadow in perfect synchrony, turns, and banks, and disappears from view.

One grackle and two red-wings remain. Why? What caused those three birds to ignore the message to fly with the crowd?

MAY 2

Great Meadow has worn white caps all day, probably in celebration of the beavers' repair job on the dam.

Both bald eagle and osprey come to play in the racing currents of air. The windspeed is perfect for the osprey, allowing it to cup its wings and move its legs into varying positions to stall and hover, its keen eyes searching for fish shapes below. Nothing. When no prey is sighted, only a slight tipping of its wings into the gale is necessary to launch the hunter up and away, only to hover at another spot.

The eagle, while an immature bird, displays eye-catching skills at high-wind flying as well. At the upper end of the Meadow, this bird tilts its expansive wings upward to catch a breeze. The wind instantly fills the feathered sails with force, lifting the bird skyward into a sideways-slipping coast. Then, as if in slow motion, the eagle sets its wings against the wind, letting its body be blown at a dizzying speed sideways from north to south across the expanse of meadow until it approaches the forest. Against that barrier, the wind meets resistance and piles up. The eagle takes advantage of the turmoil of currents, riding them up and over the crowns of the trees to begin another round of wing-dancing with the wind.

MAY 4

This late spring afternoon displays a quality of light often seen in October. In the fall, leaves provide a panorama of reflected colors. Today, spring's light reflects the pastels of birthing buds and nascent leaves, displaying a chromatic delicacy of colors every bit as beautiful as those of the fall.

MAY 5

Once again, Great Meadow's blackbirds gather and perform their concert at dusk. The liquid phrasing of their music defines sweetness. It occurs to me that I've not heard or read of any naturalist noting such reposeful bird-song. Oh, what they've missed.

These few days of warmth after last week's shivers have awakened the peeping frogs. All the songs that accumulated during their long winter of sleep are now released, one peep at a time, to every lady peeper who might be inclined to respond. I like to believe the calls have the secondary purpose of lulling this appreciative listener to sleep in total peace.

Also today: palm, pine, and yellow-rumped (myrtle) warblers in the wildlife garden that adjoins the Meadow.

MAY 9

Northern oriole returns.

MAY 10

Hummingbirds: two females.

MAY 12

The marsh is sparkling on this fine spring morning as a light wind delicately caresses its waters. Across those waters in the far cove, the graceful forms of red maples are just beginning to broadcast pink from their blossoms. Seen collectively, they lie in the valley like a cloud left behind after yesterday's sunset.

MAY 24

As I write, the angle of the sun increases and spreads its light across the crown of the new-leaved hardwood trees beyond the cove. The overall scene is an accumulation of soft images of yellowish green. A visible rebirth of green life is being played out. The forest is readying itself for a new chapter of giving, of absorbing, of producing life for all.

JUNE 2

Second thunderstorm in the young month. This one zapped an electric line someplace connected to Great Meadow. Our power went on vacation for an hour. A pair of Canada geese, who just paraded by with three goslings, did not notice that outage. Not far behind, a wood duck mother trails five ducklings. As far as I can tell, two of the three duck nesting boxes shelter hooded mergansers, hopefully cradling eggs.

JUNE 4

Every evening during the past week, I hear many amphibian throats in tune-up mode. The times before I fall asleep, and the times between sleep, are pure rapture—but the rapture is always too short lived.

Late in the evening, a barred owl's calls spar with a great horned owl's hefty hoots. This barred owl's eight-note call does not follow tradition. The pattern of notes differs in two ways from the standard rendition: first, the initial four notes of each set include a noticeable syncopation between the third and fourth notes. Second, that odd pacing is repeated again between the seventh and eighth notes, the latter phrasing being very pronounced. Will this tinkering with tradition—by adding a touch of jazz—please its intended ear? Evolution is built upon modifications that lead to success.

JUNE 7

A voiceless night. After consecutive evenings of marsh singers in full melody, tonight all is absolutely silent. Eerie. At about three o'clock in the morning, I hear a lone "harrumph" from a bullfrog. The night is cool, but only gets down to the mid-sixties. I fall back asleep thinking about reasons. I wake up without any. Rationale to explain nature is hard to come by. That's a good thing.

JUNE 9

Pied-billed grebe has been calling for about a week.

The triplet gosling family is now a twins family.

No hoodies, yet.

JUNE 12

The clear, piercing call of a pileated woodpecker stops me short, as it always does. I return to the beech tree where I'd seen a pileated drilling a few weeks ago. The bird's call invigorated my hope that nesting might be underway, and that I might have anticipated the time correctly.

Interns on Great Meadow

Two years ago, I chanced upon a pileated's active nest. I clearly recall being riveted to an activity there that has remained fixed in my brain. The question: Was what I saw a one-time action? Or might today's grand woodpeckers reaffirm what I saw?

I squeeze myself under the low branches of a hemlock tree within sight of the hole I saw in April. I have time to wait, watch, and listen. Just under an hour passes, during which time my mind wanders around and over moments granted to me by these birds.

Pileateds are not subtle birds. When they appear, they capture your attention. Under no circumstances would you look at anything else. Like some flying myth, you spy a bird, big and black as a crow, head flashing red, wings blinking white, as it streaks by and stops high on a tree trunk. Its sculptured head whips left, whips right, every move abrupt, robot-like—in a way, ghost-like. The reality registers in your eyes, yet your whole body sees it.

A call from a pileated jolts me from my reverie. Could I be so lucky? I begin to count the seconds. Ten. . .twenty. . . . The woodpecker, closer this time, calls again. Another ten- to twenty-second count. Silence.

A third call, and then everything inside me tenses up as I wait. A black arrow flies past, flashing white and aiming directly toward the nest hole. At the last split second before the bird arrives at the cavity, its partner flashes out. For a meshed moment, only inches from the hole, the two parents become one as they exchange egg-incubating duty. Only if some predator had been watching during that nick of a second would their nest site have been divulged. Such wild behaviors exist, out there to ponder.

I figure that parenting pileateds have internal clocks that regulate exchange times as eggs are incubated. First, the woodpecker out in the forest issues three calls, moving closer to the nest site with each call. (I would expect that they never use the same flight path to the nest.)

The moment the third signal is given (according to woodpecker counting), timing becomes critical. The bird sitting on the nest in the tree must know precisely when to leave. It's as if the same bird goes in and out of that nest hole at the same instant.

And now I've seen that wonder twice.

As I've mentioned, black carpenter ants are a primary food source for these woodpeckers. Any observant person walking through most types of forests will see cavity after cavity in tree trunks, each drilled out by pileateds in search of ants. Black ants forage at night outside of their protective tree. The ants chew out "rooms" of oblong cavities deep in the trees, living in clusters that can number in the thousands. The ants enter their apartments underground, crawling upward through the middle of each tree's trunk along chewed out paths. Pileated woodpeckers feed during the day when the ants are deep inside the trunks.

Question: How do pileated woodpeckers detect those colonies of ants? I have spent exciting hours observing these birds in their quest for food hidden deep in tree trunks. Here is what I have seen again and again: A woodpecker, after moving around, up, and down a tree's trunk, comes to a stop, pecks lightly on the trunk, tilts his head left, right, then moves an inch, maybe two, repeats the light pecking, listening for some sound. The procedure often goes on for several minutes.

What does the bird hear? Is he interpreting echoes from the sounds of his bill hitting the tree? Might an echo that returns after bouncing off a cluster of ants, or an ant-filled chamber, be different from an echo that returns from dense wood?

Whatever these remarkable birds determine, they do not make mistakes! Of the thousands of cavities drilled into tree trunks, an ant chamber (or several) is always perfectly centered at the end of each hole. The strength, endurance, and resolve of pileated woodpeckers is such that their excavations into trees where black carpenter ants have created large colonies can be four feet long, often two feet wide, and twelve inches or more deep.

JULY 19

I count fifty-two Canada geese on the marsh today. Mostly they show up around sunset, as if returning from their daytime jobs. They paddle en masse into the open, central parts of the Meadow where they spend the night.

Conversations range from an occasional exchange of low grumbles (those are totally pleasant) to every one of the fifty-two geese all yapping at once. Come dawn, after the daily

DUCKLING COUNTS AMONG FOUR FAMILIES OF WOOD DUCKS:

2 ducklings
5 ducklings
7 ducklings
14 ducklings

arrangements have been decided upon, they begin their deliberate, stately, slow paddle up the marsh to their hideouts in the cattail colony.

On two mornings, the parents of the goslings (and one other couple who showed up with three) joined the day's floating armada. Adjectives galore could apply to this quirky assemblage of geese, but I've misplaced my Canada goose thesaurus.

JULY 21

Two days ago, tree swallows congregated on the Meadow. For them, the excitement of migration is behind schedule this year.

JULY 24

About this year's hummingbirds: The group is comprised of three females and an over-excited male. A couple of weeks ago, I noticed that one of the two feeders I maintain for these birds remained filled with sugar water. The other was usually emptied of its sweetness every day. Why?

For a couple of days, I exchanged the feeders to see if one was preferred over another for some reason. When A was shifted to B's location, the same one was emptied. The feeder itself seemed to make the difference. I did the switch again. Same story: Feeder A was always emptied, Feeder B left mostly full.

JULY 27

Early this morning, the male bird solved the feeder problem for me: He had apparently taken personal possession of Feeder B. Any female attempting to sip from "his" feeder was immediately driven away. He obviously had spent many days on guard duty. It's interesting to note, however, that due to the females' perspicacity, they kept trying to sneak a sip while he wasn't spying. Mostly though, he would swoop down from his hideout and buzz the ladies away.

There's a bully in the neighborhood—the situation for the three females seems to have gone from bad to worse. For two days now, *both* feeders have remained full. As if one feeder was not enough to deny access to the resident ladies, now both feeders have become off limits! The publican of the local sweet bar has become a menace; a feathered dictator expanding his domain.

JULY 30

Alaska, Chukchi Sea, Russia, Aleutians. Grand sights and sightings, but the warming situation is very troubling around the world. Populations of cold weather animals are plummeting. And I return home to a drought-plagued summer.

OCTOBER 6

Far too warm: eighties late this afternoon. The facts of climate change trouble me, like a sliver in a finger, but even a few minutes at my fine home observation base suppress uncertainties that grate on my conscience.

The wild garden shrubs, my raspberries, the pear tree down on the lower terrace—all are a-wiggle with feathered ounces of migrating life. Myrtle and yellow-rumped warblers, song sparrows—all vibrating with the essence of the journey demanded of them by the tilt of the planet.

Two skeins of Canada geese come towards Great Meadow from the east. By watching over time, I've noted that these long-distance migrants alter their landing routine during hunting season. Instead of a long, slow, direct pattern to the water, October's geese come in high, out of range of lethal guns aimed their way.

These skeins circle the marsh, undoubtedly looking for any shapes or colors to be wary of. They circle a second time at a slightly lower altitude. Some uncertainty rises to meet them; some negative warning is sensed and the groups swing west over the trees. Echoes from their voices trail away.

OCTOBER 11

An immature northern harrier hunts the edges and perches upon a mid-marsh stump, inviting me to hoist my binoculars. I have not previously had an opportunity to see the soft, russet breast feathers on an adolescent harrier, so vibrant in the morning sun. The beauty of youth!

OCTOBER 29

This year the leaves began to change slowly. The colors seemed muted, but the forecasters proclaimed that the foliage would be spectacular, despite the leaves' reluctance to let go of their green. Skeptics inevitably expressed disappointment at the annual prognosticators' forecast of "a year of brilliant color."

Hey, surprise! They were in fact right. Drought conditions had apparently slowed the loss of chlorophyll, as only marginal color dragged on for two weeks. Then over a period of only a couple of days, the green faded and the annual color show came on stage.

OCTOBER 31

Recent rain has partially refilled the marsh. Canadas come through almost daily, as do a variety of ducks.

For over a week, an immature bald eagle has been worrying wildlife at the Meadow. Today, the bird is giving chase to a great blue heron, a predator-versus-prey act I'd not seen before.

Both birds are in flight. The eagle does not follow directly behind the heron, but instead keeps swinging outward in wide arcs, turning and coming toward the heron from the side. Herons fly by using steady, slow wing beats. It's plain to see that this chase is quickly taking its toll on the immature heron.

The heron heads out of sight over the beaver dam and disappears down the brook valley. The eagle follows, shutting off my view of the action. I consider the possible outcomes. While I believe the young heron couldn't possibly survive an eagle attack, it also doesn't make sense to me that an eagle would consider a heron worthy of effort. A blue heron may stand tall, but it is mighty empty of meat; at best, seventy-five percent feathers, twenty-two percent bone, three percent gristly muscle—and that mostly in its neck! But young eagles have to learn, and today it seems I'm witnessing a lesson.

Less than a half-minute later, disbelief! The heron reappears out of the forest cover and heads for the open water over the marsh. Its flight is now labored, to the point where the bird is losing what little altitude it had and collapses shortly into the marsh.

Blue herons often swim from shallow hunting spots across narrow, deep sections of water and onto shallow ground, but this bird is in trouble. Literally in water up to the neck, it can find no solid purchase for its feet close by.

The eagle is flying and watching. It takes a wide turn, and then straightens its plane of flight, heading for the heron. When the eagle swoops past with talons open, the heron can only jab weakly with its beak. The eagle turns again and repeats the striking attack three more times, each pass coming a bit closer. On the fifth pass, it drives two sets of talons into the heron's head and, with broad wings spread as wide floats, holds the heron under water. That the heron struggles to survive is obvious from the jerking of the eagle's body. Watching five minutes of the heron's death throes is unnerving.

At least fifty feet separate the eagle, burdened with ungainly prey, from a place shallow enough to gain footing and deal with its catch. It is not going to be possible to lift up and off the water with three feet of wet feathers in tow. The struggle to move the dead heron through deep water begins.

Using its powerful wings as oars, the eagle breast-strokes its way forward, stroking three, four times for each effort, resting, repeating three or four strokes, resting again, until finally coming alongside a partially submerged cluster of buttonbush still a far distance from shore.

But since the water depth has not changed, gaining the bush has not improved the situation. Though the eagle has its prey, it makes a decision:

Sublime splendor at life's end

with one mighty down-stroke of its wings, the bird lifts into the air, abandoning the heron. Alighting on a nearby stump, the bedraggled, feather-soaked young eagle spends the better part of thirty minutes preparing its feathers for flight and, I expect, regaining some energy. There is no sign of the heron. Nothing was gained in the way of nutrients for the eagle, but perhaps it learned a valuable lesson about hunting great blue herons.

Muskrat lodge

2008

Barred owls count. Cedar waxwings twitter.

FEBRUARY 24

Finally, cold that conforms to New Hampshire's traditional expectation of winter. Snow has fallen almost weekly since well before Christmas. Here in late February, three feet of snow insulate the ground.

Winter birds have been a joy all season, but I may need a subsidy to pay for feeding them. To date, they have pecked and crushed twelve twenty-five-pound bags' worth of black oil sunflower seeds. There have been well over 150 eager (and always hungry) redpolls, tiny birds that tumble out of the trees like colored pollen each morning to refill their bellies. Crows and ravens continue to play chase with any bird that gets to the peanut-butter-covered bread before them.

Heard the long-eared owl for the first time this winter. No voicing from either the great horned or barred hooters.

Rain fell some days ago during a warm spell. That was followed by a cold snap, which allowed the first inch of snow cover to freeze solid enough to support my weight. The crust gave me the freedom to hike without snowshoes. That same crust can result in the death of owls. Owls prey on squirrels (gray and red) and chipmunks, who store their food underground and can move and eat under the snow's crust, out of hearing and sight of owls. When owls face such conditions, they are sorely tested. Smaller snacks of mice have to suffice.

APRIL 2

Today, I'm just back from snowshoeing. I followed the tracks of an amorous pair of bobcats. Their four pauses in the snow for quick sex-making made my back-tracking morning a happy one. I also back-tracked a large fisher on a mission to leave his scented calling cards in a variety of obvious places. It was clear that the animal also has a future family in mind.

When tracking wildlife during stressful winter conditions, it is important to study where the animal *has been*. The signs left behind reveal its actions. If you move forward on tracks, you may well come upon the animal and cause it to run, thus forcing it to use winter energy reserves that may already be low.

APRIL 3

Tree swallows, two having returned this morning.

The record snowfall this winter is overstaying its welcome. Dense snow packs remain in the forests, and marshes, ponds, and lakes are holding onto their roofs of ice and snow. Yet red-wings and grackles, along with a starling, have followed the dictates of the longer days and journeyed north. If "first come, first served" has its advantages, I wish their lot well—but might they be pushing their luck this year?

APRIL 13

Warm weather has emitted its battle cry of heat and its forces are causing a great shrinking of winter's now retreating battalions.

APRIL 14

Ice left Great Meadow last night. Today, by midmorning, there are ducks almost wherever I look. How was the ice-out message conveyed? Twenty-five of my favorite ducks, the common mergansers, appear on Great Meadow's water, displaying a grand enthusiasm for life as only their kind can. Wood ducks are busy swimming in circles, the males in full costume vying for the uncooperative ladies. Herons are fishing, and a cadre of tree swallows swarms around the two nesting boxes. Bird spring arrived overnight.

This year's spring brought a new pair of actors to the marsh: a female bald eagle accompanied by last year's offspring. On occasion, the parent shares food, but mostly the adolescent spends its time terrorizing anything that moves or could possibly be eaten.

Off flaps the youngster in pursuit of a diving congregation of common mergansers. With talons extended, the bird attempts one awkward swoop after another but the agile, diving ducks are clearly not threatened by the game.

Practice continues: “Oh, there, a large bird with long legs and a longer neck! That might be worth a try.”

Swoop; not even close.

“Over there, a couple of beavers. Hey, let’s try a new strategy: Come in low over the water in a flanking move.”

Prey is too big.

“Okay, back to the mergansers again.”

So it might go with a young eagle. Sorting out proper prey takes awhile, and who knows how the minds of predators reason?

APRIL 18

For several days, this juvenile bald eagle has provided entertainment while learning its trade, all at the expense of anything on Great Meadow that appears to be remotely edible. (The youngster no longer wastes time on the mergansers; progress!)

Ospreys are fishing over the Meadow, but it seems that the young eagle has yet to learn the lesson that states, “When an osprey flies off with a fish, steal the fish.” The adult female eagle spends her time just watching. Bemused? Disgusted? Or exercising tolerance because she knows that patience and experience will guide her offspring?

MAY 14

Today, I watch a broad-wing hawk move its head from side to side as it seems to adjust its vision to the location and movement of some prey. With those facts processed, the hawk drops from its high perch and plummets at a steep angle, crashing into a thicket of brambles just over the edge of my deck. Its talons find their mark, and the impact has a wild result. A large milk snake is the prey, prey that is agile, strong, and now fighting for life. The hawk’s wings are spread slightly, one at an angle in thick brush and the other thrust into a tangle of thorny blackberry canes, each holding firm against the snake’s violent coiling and writhing.

Despite coils of snake wrapped around its body, the hawk drives its lethal beak into the reptile’s back. In the same motion, the hawk twists its beak, tearing chunks of flesh out of the snake, yet the violent struggle continues for at least three minutes. Watching, it's hard to understand how the bird's wings could withstand the beating they're taking, as they continue to flail against the stakes of thorns in attempts to counter the reptile's coiling actions.

After repeated thrusts of the beak, I expect the serpent's backbone has been severed, because its coils seem to loosen. Now, with the snake paralyzed, the next obstacle for the broad-wing is to pull away the coils and drag the heavy body through the thick, thorn-studded brambles to an area open enough to allow a getaway flight. With one talon still locked deeply in the thick, muscular body of the snake, the hawk must somehow maneuver out of the confining brush. I watch in awe as the bird uses its battered wings as oars, bracing each against blackberry canes, thrusting *backward* little by little! With each laborious inch gained, it pulls the snake along, eyeing a small clearing some five feet away. Once there, it doesn't hesitate. Showing amazing strength, the broad-wing spreads its battered wings and is up and away across the Meadow, the four-foot snake dangling from one foot.

MAY 15

During martini time, Runty, the resident porcupine, ambles by, making his way around and over his backyard fortress of great boulders. A small remnant of a fallen branch becomes an object of interest for the prickly one. He straddles the stick, then proceeds to undulate his body in a way that I just have to watch.

He holds his tail high while his legs, both front and hind, sort of paddle beside the log for a minute or two. All the while he nods his head, down, up, down, up. Who wouldn't be fascinated? When the porcupine stops peddling, he seems to regain his composure. He leaves the branch and waddles up onto a boulder, where he sits upright, exposing his stomach. There IT is, in plain, engrossing view—the porcupine's penis! I've watched lots of porcupines in my life, but have never given their sexual organs much thought, much less expected to see them.

Okay, there it is, erect, light pink, about three inches long, the shaft having two distinct diameters. This animal aroused himself during his log-straddling movements, and now, in an upright position and resting back on his haunches, he vibrates the quill-free inner parts of his hind legs with both front paws. With his head thrown back, he continues that self-satisfying action for some time before settling back down into a four-footed stance and ambling on his way.

I raise my martini on his behalf.

JUNE 30

Just after five o'clock in the morning, as usual, sounds from Great Meadow offer a relaxing start for the day before I even get out of bed. Green frogs "gulp" and "chucker," a mallard has reasons to quack (nonstop), a great blue heron squawks with typical discontent– but one birdcall reaches old ears as an important interruption: the always haunting "cow, cow, cow, cow, cow" of the pied-billed grebe.

Somewhere along the far shore, in cattails perhaps, in aquatic grasses maybe, this uncommon, endangered, extremely shy wetland fish-, frog-, snail-, and insect-eater is calling. This species has always been shy and always scarce, but now its numbers have dwindled to counts that are alarmingly low. Because of its rarity, the time and place of its calls have become indelible in the ears of most serious listeners.

On rare occasions, I have seen grebes here at Great Meadow, the most memorable some years ago now: a female with five grebelets in tow; the only time they have nested here. This male is late. What is the chance of his calls being rewarded?

JULY 2

Very slow beginning to the year for hatchlings on Great Meadow. No hooded mergansers, but eventually wood ducks appear trailing fluffs of ducklings (one with *fifteen* tiny paddlers!), and Canadas parade their successes forth and back across the Meadow. Their broods are all few in number, seven being the largest.

One comes to expect the unexpected when Canada geese are the subject. This year, the anomaly is a three-parent affair, with five goslings constantly hitched to three adults. I never notice favorites, as each adult is attentive and none shows the slightest irritability toward another. I expect one of the adult attendants may be an aunt.

On the other end of the spectrum of variable family relationships, there is a three-gosling group being homeschooled by a single parent.

AUGUST 15

A grebe calls briefly. Why? It is nearly September. What is known about this unique species anyway? Great Meadow has had many vacancies this summer.

I have many miles to travel to Ushuaia, Argentina, so for the next weeks, Great Meadow will be in charge of it own special moments.

NOVEMBER 3

Many years before I built our home here at Great Meadow, and up until this year, I've introduced scores of students to its natural wonders, understanding well that it has been the marsh's wildlife that registered high on their interest meters.

Turtles and fish have been caught, studied, and released. Aquatic insects, wonders that they are in every stage of their lives, have widened eyes and tweaked minds to inquire further. The Meadow's plants, stunning in their diversity, beauty, and form, have been photographed and sketched again and again, while habitats—some dark, some slimy, others muddy and smelly—have been discussed, always with purposeful thought given to the role each may play in the marsh's story.

Canoes tip over, iPhones get soaked, cameras are ruined, feet become mired in muck—but laughter always cured the problem, at least temporarily—and to the person, no one doubted the lasting value of every experience and shared idea.

Today, I've realized that a different attitude towards Great Meadow has to be confronted. A trapper of furbearing animals has arrived to practice his skills. Traps will be set and special perfume mixtures—at times from the glands of beavers, at times secret concoctions—will be applied to the traps to attract the furbearers: mink, beavers, muskrats, and river otters.

I understand trapping is a specialized skill, and one that may be necessary under certain circumstances having to do with wildlife management. Yet as one who utilizes the rich wildlife presence here at Great Meadow to promote a lasting appreciation in students (students who in today's society of electronic gadgetry are losing contact with anything natural), I have difficulty justifying any belief that trapping provides a major benefit to the residents of New Hampshire (as its fish and game publications suggest).

Wildlife comes to food. Great Meadow is a beaver-created wetland. Wetlands are nature's exemplary supermarkets and a broad cross-section of species comes to eat. Birds of prey shop for fish, frogs, and snakes. Swallows shop for offerings of dragonflies, mosquitoes, and perhaps fishflies—but they, themselves, must get in line with the frogs and snakes and toads.

All the while, the market's underwater realm—with its food-seeking fishes, worms, nymphs, carnivorous insects, zooplankton, and other life forms uncountable—shops 24/7 as they answer to hunger's urgings for growth.

This past spring, summer, and fall at Great Meadow provided, repeatedly, indelible glimpses of river otters. Without doubt, that peek into the lives of those charismatic, sleek furbearers lingers in the memories of each of the interns who were in the right place at the right time. Those otters were dependable shoppers at Great Meadow's lunch counters.

Dependable from the trapper's point of view are the habits of river otters. Otter families establish a clearly defined territory, one that can cover several square miles. They mark the territory boundaries with mounds of muddy debris, and mark those mounds with scat and musk. During the course of their playful, carefree lives, they patrol the boundaries with some regularity, adding fresh scent as they pass.

Like river otters everywhere, Great Meadow's current family members are ever curious, and when they discover any new scent while moving through their playground, they investigate it. The local trapper will have studied his area of interest. He will have noticed the calling signs left by whatever species of animal he seeks to trap. (River otters have fur that in

Summer green takes over Great Meadow

some societies is desired wearing apparel, so the pelts will bring a few dollars.) The trapper may have noted that river otter are here and today, I expect, his traps have been skillfully set, hidden, and baited.

I know what not to expect at Great Meadow next year.

I'm well acquainted with this Meadow. I have never seen an overpopulation of any species of furbearer here, so I question the need to cull their numbers in the name of "management." Overpopulation does occur, and when that happens, I would expect the appropriate agencies to weigh their decisions carefully, based upon accurate field assessments of damage, problems, or threats related to the species involved.

Beavers, fishers, gray foxes, muskrats, opossums, river otters, red foxes, minks, raccoons, skunks, weasels, and coyotes are all here at Great Meadow. Each of these species faces traps during their lives, many for six of every year's twelve months. In New Hampshire, there are roughly six hundred licensed trappers. One local Annual Trapper's Report submitted to the state shows the kill for one season:

* Beaver: 8
* Coyote: 11
* Fisher: 6
* Mink: 5
* Muskrat: 13
* River otter: 6

Clearly, furbearers face uncertain lives, lives increasingly made more difficult by human infringement and lack of appreciation, both of which become greater every day.

When the time for mentoring comes again next month or next spring, how will I respond to the questions: "Are there otters here at Great Meadow?" or "Might we see a mink?" Or from a student already well versed in ecology: "Why are eastern coyotes trapped and so often shot?"

To the first question, I can always resort to old moments, when otters played into the hearts of so many of yesterday's students. "Oh, we might catch a glimpse. Let's hope." True for the mink, too—then I'll go on to some distraction.

The discussion involving the coyotes will be more difficult to explain. A student posing such a question probably already knows that the coyote, here in New Hampshire and around the region, is the apex predator—the top dog, so to speak. That student has probably also heard of the ever-increasing complaints from foresters about problems with deer over-browsing valuable saplings, and from both commercial and private gardeners when deer do major damage to crops and beloved posies.

Eastern coyotes are efficient at culling deer. In New Hampshire, coyotes may be trapped during six months of every year and because they are often considered varmints, or otherwise nuisance animals, a complaint from a landowner may well result in yet another dead coyote.

So in fact, I know the question before it will ever be asked: "Wouldn't the eastern coyotes be a significant asset to the deer management problems if they were given reasonable freedoms to be the predators they are, on behalf of what is best for the environment instead of the shifting desires of men?"

2009

Tree frogs trill. Orioles chime.

MARCH 17

Red-wing blackbird.

APRIL 7

It is apparent that my life is speeding by full bore, with age's unrelenting hand on the throttle. For the first time, the sense of aging has become uncomfortable and difficult to ignore. The moments I've not written down on both sides of the 2009 New Year are unsettling examples of the relentlessness of getting old. Truth be known, seldom does a day go by without some event worth a word or two—but now, I've left a gap of months!

My pen should have described 2009's fine winter, should have spelled out in some detail a paragraph about the ice storm of the century, should have remarked about the many days when Great Meadow's snowy surface was sculpted in ever-changing white dunes.

But, with the Meadow's lack of activity throughout the fall, the days slipped away, time raced on, calendar pages were discarded, and recordings of special sightings were lost. I accept the ever-narrowing of my time, but not without some resolution to do better.

APRIL 23

The music of tiny aquatic frogs is cheerful, making bedtime delightful in yet another year, regardless of age. Even though I miss many of their callings, their cheerfulness is present throughout the night, bringing smiles during frequent tosses and turns.

While the joyful ardor of those frogs spills across the nearby land, the general community of life over Great Meadow is somber. The essence of spring comes with the birds, but the birds come in fewer and fewer numbers. The *whole* of the marsh is diminishing in productive scope, in quantity and quality. The community has fewer parts—fewer great blue herons patrolling, fewer tree swallows cavorting, fewer common mergansers reverse-periscoping, fewer Canada geese holding their own during verbal contests no matter the hour or time of season.

So missed are the moments occasionally provided by otters, as are the passing glimpses of the Meadow's complement of mink. The days of introducing interns to the wonders of these two species have mostly been lost to trappers. Again I question the benefits of that practice.

An incongruous thought, one insistent and not easily discarded: My time here watching Great Meadow demands that I think about the internal wisdom in operation since the wetland came to be. Descriptive words are available for man and animal that define certain symptoms of aging and/or ill health—but how does one define (or defend) the successive weakening of a natural entity?

The marsh I watch remains vibrant and ready to serve its annual migrants, but "conditions" from elsewhere, whether in the form of pollutants and careless use of herbicides and pesticides, or the developmental destruction of stop-over feeding places used during migration, or climate change, or whatever, those combinations are causing species numbers to plummet at this Meadow, especially those of its birds.

Will humans ever realize they cannot continue to make the planet's natural resources and wonders subservient to their needs without doing untold damage to all life, including their own? A great extinction is happening, but few notice because man has displaced his understanding of where heaven really is. Talk to most people and they will say they admire wildlife—but what "they" have made obvious, in most cases, is that their caring holds for only as long as wildlife conservation doesn't interfere with any plans they may have.

APRIL 24

Canada geese have long frequented Great Meadow. Many have been the nights when unseen numbers of geese gathered along the near shore and held long conversations, some raucous, some soothing.

This year for the first time, a pair has constructed a nest within a hundred yards of my watching place. With this couple close by, I listen more carefully to their verbal intercourse, all the while watching their body positions. With all the abrupt head nodding and neck twisting, one could reasonably believe there is seriousness to these conversations. And serious they are, because over the past three days, the hen has begun to lay her eggs—so all seems well.

APRIL 25

Incubation begins on the goose's nest.

APRIL 26

Martini, snack, and binoculars in hand, I record these items this evening:

* 80°F and holding steady.
* The goose has been sitting on her nest all day.
* Exquisite scenery.
* Little else going on.

The goose begins to stretch her neck beyond the nest, bringing in a variety of dry detritus and packing it around her body. Joining her on the nest are two painted turtles. When the hen stands and slides off her nest for a drink of water, she deftly avoids disturbing her shell-enclosed attendants. Thirsty, she drinks repeatedly.

When she returns to her nest, again she steps around the turtles. A couple of minutes later, she raises herself up on her nest to stretch her wings and her gander flies in, skidding to a water-spraying stop about one hundred feet away.

The gander is totally silent for one minute, and then emits a single yodel. Silence again. Following the second yodel, the female returns his call and the male instantly flushes and lands within inches of his mate.

The hen leaves her nest. The geese greet each other with vigorous head shaking and sharp neck twisting, but never touching. Greetings over, the female leads in a swim and the gander follows. She stops frequently, pulling on emerging spears of green aquatic grasses. The gander eats, but only sporadically. He is on guard.

This relaxation-and-feeding respite for the hen continues for about twenty minutes. As the hen turns back toward the nest, the male begins to engage in some head dunking. He makes a big, splashing deal out of the action. The female turns, and when he draws next to her, she follows suit and begins to imitate the dunking with the same intensity. Water flies everywhere.

The rate of dunking increases. The male draws up and pecks the hen's back twice, then, with a flourish, mounts her. The sexual stint lasts a brief five to seven seconds. Both birds follow this engagement by exhibiting a fanfare of wing flapping; twice each. Apparently, sexual satisfaction during egg hatching is an accepted behavior.

The hen turns again and swims directly to her nest. Once atop the pile (the turtles have since gone), she attends to an itch, arranges improperly aligned feathers on both sides of her body, and settles with a wiggle upon her clutch. The gander remains close by, alert.

APRIL 28

Not an easy day for the nesting goose. Temperature close to 100°F; another far too regular weather anomaly.

Indeed, an hour later, the weather changes with the force of a slap. The wind quickly stirs, picks up velocity, and, like a giant fan, forces the uncomfortable heat mass across the marsh, leaving disciplined rows of whitecaps in its wake. In less than thirty minutes, my thermometer careers downward, dropping from 100F° to 65F°! I am shivering by bedtime, but I wager there is a more comfortable goose at rest on her eggs.

APRIL 29

I take a midday peek at the nesting goose. She's on the nest with more company: three painted turtles.

The turtles sun within inches of the goose. What might be the form of recognition between goose and painted turtle? Certainly it would be different if a snapping turtle visited the nest. Whereas one is known as a possible threat, the other is accepted completely. Surely, there's more to it? Do they transmit any form of communication? Is some form of natural intuition involved? What about scent?

A late day peek at the nest reveals that, on one part of the spoon-shaped nest, more company has arrived: Two wood ducks have joined the turtles. Might the company break up the long days for the incubating goose? Does the company provide a sense of security for her? As I watch, she often tucks her beak under her wing and naps. Turtles have excellent eyesight, and at

Canada geese on a foggy morning

the slightest hint of danger they scramble into the water with a plop. That activity certainly would alarm the snoozing goose.

While I watch, the standby male exercises his protective role. An American bittern chances by, landing in a clump of buttonbush. This is considered a threat. With great water-slapping wing beats, and with neck fully extended in challenge, the gander sends the intruder flapping away.

MAY 1

The first gray tree frog of 2009 trills. . . .

MAY 2

A male northern oriole proclaims his place with rich whistles.

MAY 8

Male and female hummingbirds return.

MAY 9

A new bird arrives this evening at Great Meadow: a purple gallinule, a male, swimming at first sighting, then spending the balance of his time on a shallow, aquatic grass island, seeming to pick at many small insects or perhaps little pond snails. He pulls up lengths of grass, and then seems to peck at whatever he's dislodged. Certainly a handsome bird, bearing sharp, colorful lengths of bright orange on his beak; exciting to observe.

MAY 14

Beginning of decade number nine for me.

MAY 15

In the past, the eye-openers of spring have usually been framed in four images; the return of red-wing blackbirds, the voiced arrival of "my" pair of Canada geese, the coming of the common mergansers, and the happy sightings of otters.

This year, the red-wings came, the geese brought relaxation (while only numbering three), the common mergansers flew in, and I'm still harboring a halting hope for a glance at otters.

MAY 16

The mother Canada is still acting like a slow-cooking oven for an unknown number of baking eggs.

Early this morning the first Great Meadow goslings of 2009 swim by: a male, followed by five goslings paddling furiously to keep up, all followed by the female. This family may be three or four days old now, as they are already following their parents' examples of nipping away at shoots of marsh grasses: homeschooling, nature's way.

The watching continues with our goose nesters. This afternoon, again, the female spends considerable time tucking bits and chunks of plant material around her body as she sits on her nest. Usually, the male stays either out of sight or some distance away, but today he remains only a couple of feet away from his lady.

By now, as hatch time nears, I expect peeps are getting ready to emerge from the developing embryos inside the eggs, encouraged by repeated callings from both parents. Two-way signals and family familiarization has been underway for some time.

Grey tree frog

MAY 20

The three common mergansers are not on the Meadow this morning. The trio has never had any additional drop-by company join them in their stop-over.

MAY 21

Two days of song-filled nights and other marsh delights come from Great Meadow's many throats.

The purity of the trilling calls of the gray tree frogs is wondrous. To my ears, all the nuances of peeps and trills sound the same, but not to female frogs. The individual calls must be unique: otherwise, females would have no choices. They listen for a difference that appeals to them alone, choosing from qualities they find to be attractive, both in sound or color. I expect with the dozen or so tree frogs calling tonight, each throat is sending out vibrations unlike any other's.

It's the differences in nature that are constantly being tested. Those differences bring modifications in species, adaptations necessary to keep pace with changes in the external environments that affect lives. Last evening's bedtime air was crowded with frog music. I can only imagine having the discerning hearing of a frog.

MAY 22

The long-nesting goose is still sitting. Her eggs must have thick shells, because she seems to be well past the average incubation time for her species. Her patient, ever-at-her-call mate, remains vigilant.

A wood duck female is trailing fourteen miniature paddling machines. Last evening, she brought them to a tiny grass island with barely space enough to hold all the ducklings—but the order was given: "Get on the island and stay there until I return."

Watching those tiny ducklings wiggle themselves together on that wee, muddy patch of grass was like watching pure theater. What happened next popped my eyebrows skyward.

Once her charges were stilled, the mother wood duck flew off out of sight towards the Meadow's south cove. The ducklings, though all wiggly in their glued-together clump, stayed put. I watched and wondered. Young ducklings are normally always tethered to their mothers. But this parent was doing something different.

Ten minutes passed before the mother braked to a skidding stop just short of her ducklings. In an instant, the yellow tinged puffs untangled themselves and slid off their island of detention. As one, they tumbled away to greet Mom, seeming to shake with relief to be back beside their safety net.

What urgent calling had the parent duck answered to, in order to have left her brood?

MAY 28

Midmorning. At last (this proxy parent being more anxious than the geese he is watching), two goslings, each struggling to find use for their legs, are stumbling around the nest site. Their ever-attentive father stands alongside his mate on the nest. How many others are banging away at the wall of their shells to find freedom?

The wood duck family's duckling count is down to eleven; three ducklings must have made snacks for hungry predators.

MAY 29

Late afternoon. The goose's nest is abandoned. The result after thirty-some days: three goslings. Five of the eggs did not hatch.

Last night, cold and heavy rain; today, cold and constant downpours.

Many of Great Meadow's families are trying to cope with this test of weather.

JUNE 8

Known Canada goose count on Great Meadow to date:

* "Our" family with three goslings.
* Three families with one gosling each.
* One family with four goslings.

This is by far, the lowest count since 2000. Considering almost constant rain and low temperatures have now run into three weeks, I hope weather is the only reason for the low rate of reproduction.

On the other hand, several wood duck families have been successful: one with seventeen! I also spot three families of hooded mergansers herding good counts of ducklings.

Tree frogs continue to be the favored musicians of the spring, performing longer this year than remembered from years past. Green frogs "plunk" and "gulp," indicating healthy numbers. Audible bullfrog throats are few to date.

JUNE 9

Again, turtles struggle to deposit eggs anywhere around the open spaces at my home. Watching them dig over and over, failing every time, is always lamentable.

JUNE 10

Early this morning, the presence of a female hooded merganser asks for my attention. She swims erratically, moving in, around, and behind several clusters of bushes. The nesting boxes that she may be looking for aren't there. Because they had been flooded and broods recently destroyed, I moved them, attaching them to large oak trees between my home and the Meadow. I expect she has already spotted the boxes, but will she investigate?

For two weeks now, just after breakfast time, I've observed the single female hoodie afloat on the marsh, always in the same place. She hardly moves, always facing the same nesting box high on the hill. When she seems to become a bit restless (or perhaps indecisive), there will be a slight, quick turn, first one way, then back.

From that simple move, I know that within moments, she will rocket into the air and begin a series of high-speed circles. She heads first out over the marsh, turns back, then heads into the woods, emerging seconds later at blurred-wing speed and heads directly towards my home only to bank sharply at a diving angle, just missing the top of the nesting box. Each time, she pulls out of her slanted dive just in time to clear the water. She may repeat the clockwise pattern several times.

Upon completing whatever this-and-that she is doing, the merganser returns to her original spot on the marsh, skids to a graceful stop, executes an impressive tail wag, and resumes her original position facing the nesting box. After about nine o'clock she flies away.

I haven't the slightest explanation for her maneuvers, but it is certainly exciting to watch the flight capabilities of that small duck. I expect she will repeat all that again tomorrow.

A handsome yellow-bellied sapsucker has been calling while performing a series of careful inspections of the fine red oaks growing on the marsh side of my home. As far as I know, he must be searching for insects, not drilling tiny holes to satisfy his taste for sap.

A female mallard shows four newly hatched paddlers early this morning.

JUNE 16

I enjoy watching two crested flycatchers, both in good voice and spirited flight, as they romp among the branches of the sapsucker's oaks.

In late afternoon, several cedar waxwings come by to feed on the ripening red elderberries. I am surprised. Usually their welcomed, sleek appearance doesn't happen until late August. Yet another of the mounting anomalies? So many gears in nature's engine of life are out of sync.

JUNE 17

A male American bittern flies across a slice of open water and disappears close by in the greenery of tall aquatic grass. Only by watching the blades of the grasses sway am I able to follow that bird's movements.

He lowers his head and spears through several yards of sedge growth until he comes to a place that's apparently suitable. There, with no apparent need for stealth, he reaches forward with one of his long-toed feet and presses down, then repeats the action with the other, clearly flattening the grass into a sort of mat.

Satisfied, he then begins swaying, putting weight on first one foot and then the other. Many wading birds use a variety of stamping techniques for the purpose of irritating or flushing out prey, so I expect the activity of this bittern has a similar objective.

At the first stop, he sways steadily for roughly a minute and a half, then freezes in position with beak forward, ready to strike. No action, so back to slicing his way through the watery thicket to the next position in the grasses. Once there, he flattens another space of grass and immediately repeats his weight-shifting, body-swaying motions. Lowering his head, he poises in position again.

Strike! Something becomes food. I cannot make out what the prey is, but it appears through the binoculars to be black. The bittern does not swallow the catch immediately. Holding it in the back of his beak for about half a minute, the bird begins exercising his throat muscles, perhaps squeezing, maybe suffocating, probably crushing his catch, which eventually goes down the hatch.

POP! Up the bittern rises, clumsy as always, flies only a few yards, pulls his wings akimbo, then drops into tall grass. The stamping begins once more, followed by the swaying. He thrusts his neck and beak forward, again freezing in place. Score! Another black snack. The curtain of grass then closes, leaving me with insights perhaps few know about the American bittern.

JUNE 18

Overnight, some understanding has threaded its way among all of Great Meadow's families of geese, as they all knew to gather this morning in a pre-determined meeting place.

I watch the geese and goslings arrive from many different parts of the Meadow, assembling within a one-hundred-foot-square area where aquatic grasses grow high and tasty. The adult pairs, each with their eager, twitchy, head-bobbing charges, mingle and snap grass tips in unison. Unusual; I cannot hear any conversation.

Watching those many families side by side in a single, compatible congregation is both pleasing and sobering. The grass-picking picnic serves a major purpose for the young birds: Soon all the geese will form a skein and answer to migration's beckoning, every one fueled by Great Meadow's gifts of organic sustenance. Some of that nutrition will form into muscle, some will build memory cells that will bring the consumers back. This morning, attachments to Great Meadow have begun to form.

JUNE 21

A Blanding's turtle! This female spent the night excavating a cavity for her fertilized eggs under a rosebush alongside our guest bedroom wall. She's a large turtle, and could be the same reptile I mentioned seeing a couple of years ago.

When I see her this morning, she's covering her precious eggs. I place a metal grate across the nesting site to protect it from the long list of predators that would relish dining on turtle embryos. I am absolutely thrilled she found a spot to dig a nest, especially where all others have failed.

JUNE 30

These spring-to-summer days show little variation. Rain and gloom have dominated for two, going on three weeks. After a night of deluge and another soaking day, this evening the sun fleetingly reassures Great Meadow that it is still burning. With the momentary break in the rain and the marsh bathed in sunlight, where can I be found? Out on my deck, of course, martini in one hand and binoculars in the other, journal close by, just in case.

Small gaps in the thick clouds to the west confine the rays of sunlight, illuminating only certain sections of the marsh at any one time. My eyes automatically shift to those highlighted spots. There, framed in perfect light, a beaver is pawing globs of a favorite food into its orange-toothed mouth. When its mouth bulges full, the beaver contracts its cheeks to squeeze as much water out of the wet weeds as possible before swallowing, an action not much different from biting off a chunk of watermelon and spitting out the seeds.

A family of wood ducks moves into the spotlight. The attendant mother guides five ducklings alongside a small stump. Her instructions are clear: "Climb up onto the stump." Each youngster struggles to reach the top but there is only room for three ducklings at a time. Watching the scramble-up and tumbledown exchanges makes for totally charming entertainment. Even the sunbeam captures the comedy, remaining focused on the stump for a few more delightful moments.

Minutes later, the sun shifts its spotlight to Great Meadow's wide variety of floating and emergent greens. The strong beam focused on each plant species enhances their color, and as the sunlight moves across the water from garden to garden, my eyes are treated to shades reserved, perhaps, only for this moment. As I follow the movement of the light, a snapping turtle materializes, draped lazily over a log. After so many days of gloomy, wet weather, we both are pleased to see the sun.

I look down on two great blue herons as they cruise by on barely beating wings. My eyes catch something: a gosling, not fully feathered, alone on a spit of grass and mud. Did it paddle away from its brood? Did it get left behind due to some physical problem?

The herons fly on. A scarlet tanager whistles. A song sparrow tweedles. A squadron of hummingbirds threatens to cut short this precious time on my lookout deck, with their dashes, whiz-bys, and buzzing banks. A muskrat cuts a V through Great Meadow's dark water with a mouthful of fresh sedges; replacements for last night's mattress.

The gosling remains alone on the spit.

JULY 1

The gosling? Dead. When I check late in the day, I find it has become food for some predator. Eagle? Snapping turtle?

AUGUST 27

Thus far, this summer has failed to meet normal expectations, and the soggy conditions persist.

Again, I take great interest in watching two goose families who seem to be in charge of training goslings to fly. Getting youngsters ready for liftoff means patient teaching of Flight Instruction 101, a course that must not be failed.

It appears the two basic steps are:

1. Paddle-run across the water to gain sufficient forward momentum.
2. Flap wings faster than ever before.

By watching the adolescent goslings frantically repeat their "paddle feet, flap wings" attempts, which end again and again in awkward splashes, it's clear that the skills needed for first flight aren't programmed and must be mastered.

The parents repeatedly model the stages of flight: "Follow me, do as I do." All through these practice sessions, the stentorian instructors' encouraging calls reverberate forth and back across the marsh. No matter the family, an enthusiastic audience obviously has a positive role in a performer's success.

When I observe these youngsters running fast enough and flapping speedily enough to actually fly, I try to imagine what a feeling it must be to suddenly be airborne. Even after a week of practice, the students can be picked out of the moving skeins by their erratic lines of flight—but they are flying!

Learning to touch down on water is another matter. There is no way to avoid it. The few initial landings are more breast-splash than feet-glide. What follows these trials only adds to the entertainment value of watching these admirable animals. Victorious, congratulatory, and exuberant honking, head bobbing, neck weaving, wing flapping, and tail wagging—all are outward signs of the parents' hearty approval of lessons learned. Once a family returns to the marsh after a first flight, a swim-around follows, and all the Meadow's goose families join in the rejoicing.

Perhaps the most difficult skill for young geese to master is that of "tipping up." Apparently, much yearned-for food lies hidden underwater, at just about the length of a goose's neck when tipped down. To reach those succulents, a goose must thrust its head and neck downward, then its legs forcibly backwards, all at an upward angle. It's that angle—tipping the head, neck, and chest down into the water while pushing the tail end up to the vertical—that does the trick. Being on hand to watch such lessons is priceless.

AUGUST 30

First great egret of the year: a single. Very early.

SEPTEMBER 22

Blanding's turtle update: The required three months have passed since the mother dug her nest on June 21. As the deadline for hatching draws near, I enclose the already-protected nest site with twenty-four-inch hardware cloth to prevent any hatchlings from becoming victims of predation. If the newborn turtles survive predation in the nest—and few do—the next period of danger comes after they leave their eggs and head to the wetland. Raccoons, foxes, coyotes, fishers, minks, crows, ravens, herons, owls, even chipmunks and stray cats, would all consider a quarter-sized turtle merely a bite-sized snack, and those hazards form only a small portion of the gamut the turtles will face. Once in the water, luck must carry them past an array of hungry fish, amphibians, aquatic mammals, other turtles, and birds with lethal beaks. Every turtle, especially those species whose numbers are limited, is important.

SEPTEMBER 25

Of the six or seven regular hummingbirds feeding this summer, all left ten days ago, except one. As of this date, that lone bird is still filling his calorie tank in preparation for the trip south.

OCTOBER 1

Hunting season for waterfowl opened today. A father and his teenage son arrive at Great Meadow well before sunlight. Camouflage covers hunters and canoe. Tucked close into a barrier of buttonbush, they set their decoys and make weapons ready.

I watch through the dim light as the permitted shooting minute approaches. Having participated in similar scenarios, I identify with the expectations of the hunters. I know the hopes of the father. I know the unease, arising from the possibility of failure, that's building inside the lad. Perhaps he's hoping no birds will even show up.

Morning's faint light is partially hidden by thick curtains of drifting mist. The eyes of both father and son peer into the moving veils. Their ears listen for some signal from flying birds. The minutes feel long, made longer by the chill of the near-freezing morning air.

The father hopes for waterfowl to give credibility to his reasons for being here. His son probably holds mixed feelings. Successfully drawing ducks or geese to the decoys would make his heart leap with excitement, but if the waterfowl do come, and the boy shoots and misses, what will his dad think about him?

Having experienced it all myself, I understand about the first minutes of anticipation and nerve-tight focus. I also know that the mind begins to wander when there's still no action after the first hour. Instead of closing out the insistence of the cold, nerves begin to take command of chills, and cramps set in.

As I watch from the overlook of my home, the mists dissipate and the sun exposes all in the Meadow with its brightness. The canoe moves away from cover. The lad in the bow still holds his shotgun across his lap, but the hunt is over. Dad collects the fake ducks and returns them to a sack. As he turns the canoe shoreward, I sense their disappointment, but hope there will be commendation for a youngster's well-controlled impatience.

STRANGE. Not a single waterfowl has swum or flown by these or other hunters today. Other than the eleven Canada geese I saw on the water four days ago, no migrants have come to Great Meadow. Is the migration early or late this season? Is it because of the inclement spring? Long-time familiarity with a natural place adds gravity to these questions.

Finally, the last of summer's hummingbirds left today; a tiny speck of flying life facing an unbelievable journey. Creditable surveys have shown that "my" male ruby-throated hummer, after making his way southward along the east coast, will turn towards the Gulf states, gathering stores of fat along the way. He may then head due south and fly, nonstop, up to nine hundred miles across the Gulf of Mexico to some destination in either Panama or another place along the northern coast of South America. Talk about the efficient use of energy. . . .

OCTOBER 22

It's barely light yet and drizzly, but clear enough to see a wide and larger than usual V disturbing the smooth surface of Great Meadow. A quick peer through binoculars reveals a swimming moose, a female. She is moving from south to north, her wake dividing the expanse of the Meadow almost exactly into east and west.

She moves ever so slowly, looking right, left, and straight ahead. Only her great head and constantly swiveling large ears show. It seems she's out for a leisurely, early morning swim. Barbara and I watch for long appreciative minutes until the long-legged paddler finally disappears from view behind the branches of marsh-side trees.

As I stand by the window, deep in moose-thought, a pied-billed grebe makes a far smaller V out in the marsh as it swims towards the eastern shore.

NOVEMBER 2

Red-winged blackbird!! That species should not be here in November.

NOVEMBER 26

Thanksgiving Day

A doe, perhaps spurred on by hunters, or just trying to get from one side of Great Meadow to the other, swims from the western shore across the wetland's expanse, over to the eastern side. When she gets about halfway across, a flock of Canada geese that had been postponing migration takes interest in her.

All of the birds paddle furiously toward the deer as a group, quickly closing the gap. They escort the doe like a fleet of tugboats until she steps out of the marsh onto dry land. The doe shakes the water from her coat, and the geese turn and swim back out into the marsh.

Simple curiosity? Just something the flock decided to do to pass a few minutes? I wager there is so much more to their actions.

DECEMBER 10

At breakfast time, a large mink catches my eye. A full-moon night stilled the surface of the marsh to the point where it froze into a skim of ice. The mink is stepping, ever so gingerly, across that fragile skim lest it break. Moving along, one cautious step followed by another, is so out of character for mink, whose natural stride is to loop-step.

It approaches a newly formed muskrat lodge. The moment it steps onto the slanted edge of that structure, it's as if it had bumped a wasp's nest.

Instead of wasps, muskrats—one, two, three, then four, five, and six—pop out of their lodge and swim erratically around their dwelling through the broken ice. All the while, the mink stands atop their lodge, seemingly fascinated by the panic it has created. The muskrats' greeting holds little interest for the mink. It slides off the lodge into the ice between two muskrats, and off to some other destination.

Meanwhile, the muskrats don't go back into the lodge, perhaps figuring that while they're out and about, they might as well mill around and continue to enjoy the morning slipping and sliding around on the ice.

DECEMBER 12

While I'm standing out on the deck, two flocks of geese fly by, close enough for me to hear the whistle of the crisp morning air passing through their primary wing feathers.

Those birds represent the epitome of winged energy. I always envy flight.

Turtle tracks in mud

2010

Great egrets rasp. Great blue herons gronk.

JANUARY–FEBRUARY 2010

Off on Amazon River headwaters excursion out of Iquitos, Peru, a seaport 2,300 miles from the Atlantic Ocean. . . .

MARCH 1

. . .and like the migrating birds of last fall, the days of my life fly away. Ninety-odd days have escaped my journal and 2010's winter was spent elsewhere.

Winter, to date has been uneventful, with January's snow cover a barely adequate protective insulator for the lives underneath. One windstorm of high gusts spun through the state and played rough with trees, tossing them across power lines with the resultant inconveniences. Thankfully, my grand white oak stood, giving up only one large branch to the wind's demands.

MARCH 8

Three Canada geese flew off towards the river early this morning, cruising over the ice-bound marsh. I watched every wing flap with accompanying respect for their successful return.

Trouble across the marsh. Without notice, town conservation officials for land that abuts Great Meadow have hired someone to spray a highly toxic herbicide on and around the pastures and forested edges of the easement-protected property directly across the marsh from my home. I have been the annual steward of that easement since it gained protection in 2000. Key players with direct interest in that land, including myself, were not consulted about the spraying.

Several illegalities were committed, and I have spent much time with the town (the property owner), the local land trust (the grantee), and the state

Fog on Great Meadow

(the officiator), trying to point out the long list of procedures that were ignored. Too late to stop the toxins, and no way of assessing the long-term effects upon the pastures' flora or fauna or on Great Meadow proper.

MARCH 21

Spring! A pair of hooded mergansers are floating on one of the few patches of open water not far from my windows. During the night, the few areas that were ice-free have refrozen.

Following an early week of heavy rain, two days in the sixties have left the marsh ice in a weakened and ever-blackening state. A muskrat maneuvers his way along the near shore at noontime. Red-wings and common grackles wheeze and sputter their spring announcements, while the crows exhibit their version of spring fever by performing on-the-wing tricks in a stiff breeze.

This afternoon, thirteen common mergansers came slanting onto Great Meadow. My heart does up-ticks whenever I see these exciting waterfowl.

My anticipation of spring's arrival seems to intensify every year. Clocks tick and calendars chew up the days. The fact that life gets shorter every day goes for everyone, but when one is living in the ninth inning of life, the rules of the game change, becoming increasingly more difficult to interpret. Nowadays, I sense that the rules of my game have changed: Instead of being allotted three outs for my final inning, I may be limited to only two, so I must make the best of it before the game is over.

It is unnerving to be living during a time when, by simply watching the birds here at Great Meadow, they alone make it clear that something is askew with their climate support system. On average, the visits of even the usual species have dropped precipitously. Traditionally, when spring brings the players back to Great Meadow's stage, each winged arrival reinvigorates hope, along with the satisfaction that I've made it to another new season of life.

Hooded mergansers were few last year, and the common mergansers' time here was abbreviated. Even seeing those few arrivals today has meant pure joy. I'll be watching for more.

MARCH 22

Finally, Great Meadow has come alive with spring's wildlife early this morning.

Two beavers make wakes between the remaining islands of ice and islands of shrubs. At cocktail time last night, a royal, mature bald eagle swung by the deck, and this morning, as the beavers paddled into thick cover, the same eagle alighted on mid-marsh ice.

For the next hour, the air was busy with fly-bys of wood ducks, mallards, hoodies, red-wings, grackles, crows, and ravens. A squadron of least sandpipers flashed their under-wing white in one gravity-defying turn after another and another. The Meadow is being injected with spring's tonic. Such a fine run of days, but I still yearn to sight an otter.

MARCH 23

A second bald eagle, a three-year-old, keeps our breakfast busy with so much to see between bites. Spring is great at restoring positive emotions.

MARCH 24

An errant page from winter's story spread across most of Great Meadow's face early this morning, something to do with water becoming cold

Just at sunrise, Great Meadow's world is tinted in a single drop of water

enough to stop moving.

It got down to 18°F overnight, and the Meadow, all except its deepest, most exposed areas, froze over; a late winter's trick. Apparently, one pair of wood ducks stayed through the night, where they kept moving just enough to keep the ice from congealing around them. They were still there when the sun appeared over the southeastern hill.

Now at midmorning, the ice has mostly disappeared, shying away as ice will from the sun's unforgiving rays. The wood ducks move into the ice-free channels, the handsome males ready to swim before the critical eyes of the females.

MARCH 27

Just off shore next to a colony of newly emerging cattails, three juvenile beavers (a satisfying sight to know that offspring have been given life) paddle southward in their dawdling way. The trio passes only a few feet away from a great blue heron that just speared a yellow perch. No one takes notice.

APRIL 1

The eastern phoebe is back.

APRIL 4

A killdeer announced its return and I'm thrilled. This species ignites wonder and appreciation in me. They demonstrate, with every sweep of their wings, their power of flight.

Killdeer are the most vocal of the plover family. They sport two black rings around their necks, punctuating nicely the black and white flashes seen during flight. Fortunate is the birder who catches a glimpse of a rub of color under the tail. I implore the bird to nest nearby just so I can feel close to this exciting species.

APRIL 8

A couple hundred tree swallows drop by Great Meadow, wreaking havoc upon the newly hatched hordes of aquatic flies and midges. Watching them cut dizzying, fly-gobbling paths through the pink air of sunrise amounts to dessert before breakfast.

APRIL 14

Two weeks ago, the nesting pair of Canadas confirmed they were the same pair we watched last year. Prior to actually returning to their nest site, they spent those two weeks perusing every cranny, island, and cluster of close-by bushes, making sure of the site's acceptability and security. Much attention was then paid to refurbishing the nest according to the female's specs. On occasion, the male stands at the nest's very edge and carries out his own neck waving, constantly surveying. I don't suppose his mate pays much attention.

APRIL 17

I assume the female Canada is laying her clutch, as she does not move off the nest. Her responsibility is clear, and her guardian is alert to any threat.

So far this year, he is vigorously and effectively patrolling the entire southern half of Great Meadow. No goose had better fly, or even swim, within his staked-out territory. Those who do, mistakenly or perhaps with some other intent, better brace themselves for a winged assault. This male means business. Any considered intruder must either take flight or prepare to be blasted by fifteen pounds of attacking fury. Indeed, his lady has chosen well.

APRIL 19

This afternoon, some turmoil is underway just off shore. After a brief glance, I determine that only one event could create such a stir: The mating of snapping turtles. How much more I appreciate their actions, having witnessed the struggle on two previous occasions. I'm mostly convinced the act is an epic test in copulation. I rush down the steep embankment and push my canoe onto the Meadow to get closer to the scene.

As I've noticed before, the engagement seems violent—great clawed feet probing, pushing, and grasping while muscular, extended necks twist and curl—but I've never seen biting during the contact (although I have watched many middle-aged snapping turtles scrape and bite while making some stand). The grinding sound of the turtles' shells and toenails knocking and scratching are unduplicated in nature. With such a thick, cumbersome carapace and a stiff, armored tail, a traditional mating position seems unlikely.

These reptiles, possessing a lineage going back some 230 million years, continually fascinate me. Admittedly, while I cannot draw conclusions from three isolated incidents, I tend to believe there may be truth to the idea that snapping turtles copulate carapace-to-carapace. However, in sharing my scant observations of this mating act with professionals holding doctorates, they have made their contrariness clear.

The male eventually turns the female over and holds her in position. All action then ceases and the coupled turtles slowly sink. The wave action from their repeated engagement underwater indicates the aggressive nature of the coupling. During my current on-hand viewing, I can't be sure whether coupling with this pair has occurred or not.

The struggle continues. The patrolling Canada male swims by for a look at the commotion. He stops within twenty feet and also seems to be transfixed.

In the shallow water, I watch the female turtle's long neck extend and go completely limp. Both turtles remain motionless for two or three minutes, until a dual trail of bubbles begins to rise alongside the pair. Once the turtles surface, they separate and swim away slowly.

I sit in my canoe, quiescent. All is still. I've been given much by those reptiles.

I gingerly ease my canoe back to the shore.

APRIL 20

An otter! He's huge. He's moving fast upstream this morning, cutting the

Meadow's calm water in half with his wake. I silently beg him to stop, to fish a little, to show himself, but he quickly passes from my unblinking eyes, out of sight. I even watch the re-gathering of the waves he made, as if to seal the moment in memory.

One otter. One very large, male otter. Going where? Coming from where? Did he find any fresh scent?

APRIL 26

An ill-timed tragedy: A female hooded merganser had been laying her eggs in one of the duck boxes I had tied to the red oak that grows close to the edge of Great Meadow. From my breakfast table, I notice that the nesting box is now gone.

Upon checking, I see that the bottom brace of the box slipped from its mooring at the base of the trunk. The night brought a gale, apparently putting enough pressure against the box to dislodge the anchoring pole. When the base slipped, the entire nest crashed. Not good.

I find the female merganser apparently survived and escaped, but her eggs are scattered. Five are broken. Four were thrown out of the entrance hole and rolled far down the steep bank, although none of those broke. Three were still in the nest, but covered with yolk. A sorrowful mess.

I clean the nesting box and reattach it to the tree. I put each of the undamaged eggs back in the reassembled nest; an act of hope, although I know sincere intentions sometimes end unhappily. Lending a helpful hand may be folly.

APRIL 28

Two immature bald eagles. One osprey. In a high morning wind, the eagles are practicing their aerial idle-time before a pleased audience.

A medium-size snapping turtle, on his back, dead this morning on the shore. The eagles repeatedly swoop downward to check out the potential food, but neither bird lands for a closer look. Surprising in a way, because large snappers possess many grams of nutritious meat.

MAY 5

The pair of nesting Canada geese I watch on a daily basis demonstrate the oft-believed notion that Canadas mate for life. Watching this couple, there is no reason to not accept that warming thought. While it is well known that mates, male and female of this species, often stray in their relationships, this pairing is meeting a high standard for assumed fidelity.

The female seldom leaves her egg-warming duties. The male is on constant, wide-ranging patrol, although ducks, muskrats, beavers, and turtles can move in close proximity to the nest with impunity. As observed before, the ducks and turtles even seem to be invited to lounge on the nesting platform. Compatibility among wild creatures, at least with these species at this point in time, is not unusual; how refreshing and heartwarming to see.

MAY 7

Not every day, but often, the female goose gets off her nest. When she does, her mate notices instantly. He launches himself from his patrol station, issuing his loudest squawks and flying directly to the nest. Both birds comment loudly, and express their bond through bill-and-neck twisting, touching, and sensuous rubbing. The male stays close to his mate as she consumes bits of sustenance from tender, underwater roots and grass stems. The snack breaks are short and the hen, I expect reluctantly, always paddles back to her nest closely escorted by her champion.

Once back on the nest, the female pulls feathers from her breast, re-fluffs the nest bed, and repositions her eggs. Meantime, her mate, while remaining in the water, repeatedly stretches his neck over the nest, seemingly interested in her every move. After roughly a minute, sometimes two, she settles upon her eggs with a comforting quiver. When all seems in proper order, the male departs, resuming his role as dutiful protector. At least during the trying nesting period, this pair is truly inseparable.

MAY 8

A new wrinkle at the goose nest this morning: it seems that decoration day has arrived. Dozens of two-foot-long strands of bluejoint grass have been strung around its perimeter, completely encircling the sitting goose. Why? Which parent went to the trouble of pulling, then bringing that grass to the nest? The green border-like elaboration apparently is not food. It remains in place until late afternoon, when the male swims by and removes it.

> SPRINGTIME ACTIVITY:
>
> Killdeer seen and heard over the marsh on a regular basis
> Northern oriole back
> Phoebe nest full of eggs

MAY 9

A hatch! Five goslings trip around unsteadily on their nest, parental attentions on full display. Three unhatched eggs remain. Might these parents feel some definition of the word "proud?" To me, at least, they show swagger.

In a short while, I notice a second pair of Canadas herding a batch of seven tiny yellow swimmers ever closer to the new parents' location. Following in their wake come two other geese, seemingly unattached to goslings. All are swimming within the previous "off limits" area. The gander pays no attention. Things are different now.

MAY 12

Again, a common gallinule, early this morning. Incredible! Unexpected. This bird belongs to the family of rails, about which I know little because they are uncommon in New Hampshire.

I am treated to an unforgettable, eye-popping moment through my binoculars. This unique bird is shaped quite like a barnyard hen, but his colors challenge that comparison: a red bill (with that color running up and covering his forehead); wavy lines of pure white feathers on each side of his body, framing his wings like parentheses; his chunky body bluish-gray.

As distinctive as those characteristics are, when this bird moves, his long-toed feet dominate the moment. I watch as he literally walks on top of a bed of broad-leaved water lilies. With quick moves, he flips over lily pads, often pecking off some juicy tidbit. Apparently the food service is inadequate, because in short time he flies away to another section of the marsh; far too soon.

MAY 13

It's been several days with no sign of either set of goose parents with their young.

MAY 14

A birthday gift! Today, "my" goose parents show themselves and their five youngsters, all still alive.

Glancing away from that family, there, sitting on a boulder that is part of a terrace between my home and the marsh, is my garden nemesis, the resident woodchuck. Hanging from her mouth is a baby chuck. Perhaps moving day has arrived.

One, two, three, four quarter-grown infants are being transported to new digs. The mother carries each of the first three by a flap of skin along their throats, the fourth by the skin along his ribs. All go into a tight curl and make no moves during the mother's bouncy, stop-and-go journey up, over, and down a long series of boulders that apparently make up the path to their next home.

Being on hand to watch this family move is fortunate, but there's more to come. A good five minutes pass, the goose and her family have disappeared, and just as I turn to leave, there is the mother chuck returning once again to the old den. What could be next?

Of course! The furniture has to be moved as well. She makes two more trips, both with large mouthfuls of nesting materials that she carries from the old home to the new. Familiarity with well-known scents will provide comfort for her babes.

Now, with at least five woodchucks to contrive devious means of devouring my vegetables and flowers, I know I'll be in trouble soon. In past years, each of my "creative" strategies to thwart the furred hooligans was summarily defeated. All the green delights of last summer's garden ended up feeding the woodchucks, not the gardener.

MAY 16

A ruby-throated hummingbird, a male, is back. Just like last year, while I am at the breakfast table he hovers right outside the window and seems to peer inside. His request is self-explanatory. As his appointed butler, I comply and furnish his bar with sweet stimulants immediately.

MAY 25

A crested flycatcher sends forth his upsweeping whistles from somewhere within the oak canopy, announcing his return from places far south. When the bird flies, his tail is a showpiece—mixed pastels, pinkish/orangey. . . sort of, but not quite.

MAY 26

The past two nights have lifted wave after wave of the sounds of this new spring into our bedroom. The tree frog orchestra sends pulsing call notes into the night, bringing sleepy comfort.

MAY 27

Many have been the times I've watched the Meadow's great blue herons catch fish. Depending upon the species and the size of the prey, the technique of swallowing differs, especially when the prey is a horned pout.

I have previously written about the step-by-step cautions the heron must take to avoid strangulation from that fish's spiked fins. Yesterday, I watched a mature heron yank a horned pout out of the water that was a good twelve inches long. Either this bird had delivered a paralyzing death strike with the initial thrust of its sharp beak, or some other skill was at work.

The pout caught, the predator walked the prey to the nearest mudflat and laid the fish on the surface. One, two, three, beak stabs, and the heron picked up the fish, deftly juggling it into the headfirst swallowing position. With its head now pointed skyward, the heron stimulated its throat muscles into the necessary pulling action, and the large meal bulged down toward its final destination.

Usually, swallowing such a horned fish takes many minutes of pre-swallowing preparation, but this skilled heron dispatched its catch in less than one minute; experience at work indeed.

MAY 28

A green heron, a male, this morning. This is a crow-sized heron, and finding the "green" described in his name isn't easy. A faint splash of color shows on his legs, but is hard to justify. Even his body feathers tend to be more bluish than green.

This little bird flutters in close by and immediately begins hunting in the grass of a mudflat, stalking prey with the utmost deliberation. He freezes in position, holding that pose for a minute or more. Stab! A four-inch fish, a horned pout. When the sleek, little heron flies to a different grassy location, the rest of the performance is hidden completely, taking away my opportunity to learn more about the fascinating technique that species uses to swallow spiny fish.

JUNE 1

The evening brings a Blanding's turtle! I assume the same female that visited last year is answering to her responsibility to deposit the eggs she is carrying. There she goes, marching across the hard stone pack of the driveway, passing up the nesting location in the flowerbed that was used last year, continuing on, finding no place of satisfaction. A thrill to see that turtle again.

It is seldom good to interfere with wildlife, but knowing how the circumstances in my yard affect turtles, I pick her up and set her down in the flowerbeds close to the forest, over where the soil is sandy. This Blanding's is not shy like others of her species I have held. Each of those would pull in their legs and hide their entire head behind their hinged plastron.

Now on the ground, this turtle paces quickly ahead, pausing repeatedly with her head held high. She looks right, left, right, left, seeming to reset her bearings. Decision made. Off she plods, away from the flower garden, avoiding one boulder after another and heading west into the woods. I don't expect to see her again.

JUNE 2

Surprisingly, this afternoon I spot the Blanding's crossing my driveway, returning, I hope, from her nesting place somewhere in the woods and heading for Great Meadow. I follow as she makes her way between boulders to the Meadow's edge. I fully expect her to scramble away and disappear under the water's blanket of duckweed.

As noted, however, this turtle is not shy. Even though I am close by, she does not hurry away. She slowly wades through a reach of muck, stops, and raises her head with one eye aimed in my direction. More than three minutes pass. I babble words expressing my appreciation for seeing her again. Disinterested, she moves into an inch or so of algae-filled water, but then stops, and turns her gaze toward me again. Fifteen minutes pass before she slips away into her Great Meadow habitat and out of sight.

An encounter. A meeting. Might we have exchanged some unrecognizable communication? I pause over such moments.

JUNE 3

Snapping turtles keep appearing in my yard on egg-laying missions. Very disconcerting to find that during the night, yet another female has scaled the steep embankment searching for an acceptable nesting site, with no success. But I believe I may know the reason they return.

When a watery habitat is surrounded mostly by forest, female turtles seek any sunlit opening along the shoreline. In those places, the soil may be such that they can dig cavities and deposit eggs. When the meadow-side lot was cleared of some of its many trees for my home, sunlight suddenly flooded through the clearing. The turtles noticed—but the opportunity created was deceptive. After climbing the bank, only stones and hard-packed gravel await the turtles at the top, a man-made, false environment. In the flowerbeds, any underlying soil is covered with mulch. To turtles, that is unacceptable. Occasionally, the soil sniffing returns a favorable response and digging might begin, but stones wait just under the surfaces and the turtles plod on, searching, searching. As a result, the areas around my homestead are pockmarked with failed holes.

As the years have passed, the number of turtles trekking to areas around my home intending to dig nests has dropped substantially. It hurts to watch the physicality exerted by these multiple failures. As I watch, the female turtles give up and trundle back to the marsh each year. Again I wonder: Are the eggs wasted? Might one assume the eggs are slowly absorbed into the body of the female from whence they came?

JUNE 10

Cedar waxwings. Far too early, but so are the red elderberries they are feeding on.

With these two species, one might point to a positive adjustment to climate change—but the example is only one, and a limited one at that. How many species can, across the spectrum of wildlife, successfully readjust their habits to rapidly changing conditions?

JULY 19

A single great egret!

JULY 22

Six great egrets! To date, three has been the record.

This summer's drought, so the experts say, will surpass any yet recorded. Add record heat, and Great Meadow's volume of water is shrinking fast. Exceptions in nature (like this drought), challenge some aquatic species while others benefit. Because broad areas of marsh bottom (mud) are now exposed, many species of birds can find snacks not otherwise available. For example, Great Meadow's hatch of four killdeer are constantly flitting, landing, and dashing forth and back across the flats; a fling of spotted sandpipers enthusiastically snaps up nutrients, wheeling by on wings in perfect formation toward the next supply of snacks. It's just the right type of environment for these birds.

JULY 23

The morning's contrasting colors—greens from clusters of leatherleaf and buttonbush, and browns reflecting from the mud—create an unnaturally flat wash across the dry Meadow. Evening's light brings the opposite: the islands of green become isolated, their colors sharpening and darkening, while the mudflats deepen to an intense brown, making me think of photos of Earth as viewed from space. I imagine outlines of continents surrounded by seas.

This morning, tree swallows moved across the scene like scrambling fighter jets, grabbing flying protein for their impending journey south. One struggles when thinking about the forced adjustments they may have to make across the spectrum of this one Meadow as the world's climate itself, due to an overload of CO^2, is forced to change.

JULY 24

Drought and heat continue to impose their burdens. The six great egrets remain extremely active. Apparently, they are finding ample food, affording them the energy to fly here and there around the circuit of Great Meadow, giving chase to each other or simply cruising around, exploring new niches that might offer snacks.

In watching, I've noticed differences between the egret's food-stalking style and that of the great blues. The egrets don't display nearly the same amount of unlimited patience that is always evident with the larger herons.

The egrets step nervously, seldom striking a steady pose, repeatedly stabbing at targets as they hunt. Unlike the herons, who move with utmost stealth when hunting in any depth of water, an egret lifts each foot *above* the water's surface, and then plops it down again. A heron seldom splashes when prowling for food, instead folding its long toes tightly together and then moving its whole foot and lower leg slowly forward in the water, causing as little disturbance as possible.

I suppose whether stealth or blunder-ahead methods are employed, what works is the end result. To each his own. What is important is that life has given two magnificent birds to this planet, and to this meadow.

JULY 27

I look out over a birder's paradise (truer today than yesterday) from my home's high windows. Great Meadow is busy with many birds this early morning. Again tree swallows, more having arrived from northern climes, swirl and busy the marsh's air, interweaving with the actions of all the other birds that are also moving on and over the drought-stricken Meadow.

During my years of watching, Great Meadow has not had more than one pair of killdeer. Today, I count eleven birds. Among those I see several adolescents. All are busy inspecting the muddy dining table for tidbits. I suspect the additional numbers of killdeer arrived from elsewhere upon learning of the awaiting banquet of Great Meadow.

A semipalmated sandpiper flashes, dips, and quick-turns his speedy way through the thickets of swarming tree swallows. Meanwhile, the annual flying lessons of tentative young Canada geese become enjoyable distractions. I smile as I watch.

The occasional common grackle or red-wing flits by in eagerness to join the growing cast of hundreds, as migration fever grows.

The six great egrets are constantly active, thus offering—really inviting—anyone interested in watching to see how they conduct themselves. Over a martini, at the final sipping hour, I learn a bit more.

Despite their likeness to great blue herons, although half the size, the egret's style of hunting is very different. Their leg/foot action I've mentioned previously; they also use their supple necks in ways completely different from great blues. When the larger herons zero in on prey, their heads and necks are absolutely rigid, save for the almost imperceptible forward-downward, get-closer-to-prey-before-striking movement. Stalking of a single prey may go on for several minutes before success or failure ends it.

Egrets have little tolerance for stillness. They stroll along and anything that moves or looks appealing receives the business end of their sharp beaks. When their attention is captured and they do halt, they flex their necks and heads (not unlike a hawk or an owl), trying to focus on the meal at hand. The side-to-side wavering is rapid and usually ends with a strike. Like a chicken pecking at bugs, egrets often jab at prey while wading along. They stab tiny fish and small frogs, positioning them quickly in their throats before muscling them down their gullets. If great blue herons tried to live on such small fare, they would quickly starve. Watching egrets in action requires far less patience.

JULY 28

Strange. A great blue seems to be performing a version of a quickstep. As he stamps left-right, left-right, his body spins 180° one way then 180° back, his wings extended outward half way. I watch this action for just short of a minute—then a dark thought jabs into my mind.

Occasionally, there will be fishermen on Great Meadow. Because the heron seems a bit frantic, it occurs to me that his feet may have become entangled in monofilament fishing line. The dance steps—the spinning, partial wing flapping, snapping head moves, and beak stabbing—is far from natural.

Ah! Mystery solved: The heron is holding a medium-sized common water snake in his feet! He quickly pecks at the snake, then jerks his head upward, dances again, and repeats those gyrations. I'm quite certain this is a learned behavior to avoid the striking snake. Good show, and a relief to dismiss the thought of fishing line.

At noon a large shape moves between two thickets of buttonbush, those thickets now growing as islands set in lengths of mudflats due to the drought. When that shape appears a second time, I recognize its rear-end: eastern coyote. This is certainly an animal in fine fettle—a male with a full tail and sleek brown hair mixing to darker brown along his back.

Across the marsh, the coyote takes a few purposeful steps and stops, constantly looking in the direction of my home. When nothing arouses his suspicions, he ignores the house and continues on his way. When he does stop, it's always behind some greenery.

His nose seems to be tugging him along over grassy clumps and through stretches of mud, in which he often sinks to his knees. Soon enough, it becomes clear he's not receiving false messages, as I watch him stop and pull a fur-covered snarl out of the mud. No bones are attached to the catch, and even the skin looks rotted. Nevertheless, the coyote has come a good distance to reach this scent.

His first act is to rub his neck and shoulders into what has to be fetid flesh. Then, he gives his muzzle a thorough application of the same odor. For good measure, he follows up with a complete body roll in the much-desired stench.

With that pleasure attained, the coyote turns his attentions to the edible portions of the treasure. I have to believe this is a very hungry coyote, eating the food from necessity. Each bite is accompanied by rapid head-shaking, a sign I take as him saying, "Hey, this is a pretty lousy tasting lunch."

Perhaps feeling a tad uncomfortable by the proximity of my home, the coyote gives the property one final glance, turns, trots into a thicket and disappears from sight.

JULY 30

Seven great egrets! These active birds sketch dazzling lines of white above Great Meadow each day; they are exquisite attractions for a backyard.

AUGUST 10

During this prolonged drought, when Great Meadow has receded to less than a quarter of its regular breadth, finding food for many wild species has become simply a matter of showing up. (How they learn that a feast table has been set to their satisfaction is their secret.) With food so plentiful, bellies are quickly satisfied, so the various species of wildlife spend their spare time giving chase, usually to each other.

Something new again this forenoon: great egrets vocalizing. I've never heard them make a sound in previous years. This year's "all bellies are full" free time has resulted in attention-grabbing egret talk. Unlike the witch-like squawks of great blues, the egrets issue sharp yelps while giving chase. It is not really clear if the cries come from the chaser or the chased. Their utterances are more like muffled crow calls, with the "caw" drawn out a smidge longer, as if they were choking on the sound.

AUGUST 16

The drying sediment under Great Meadow continues to drink what small amount of water is left on its surface. Clouds darken and offer false promise of moisture, but withhold their precious loads for other meadows.

As the egrets fly forth and back across the Meadow, carrying on some mission of the moment, underneath, the drying sediment keeps absorbing water as it and its inhabitants continue to adjust to the drought.

The Meadow's fishes lose ever more space, while its amphibians hunker down in muck or take refuge in protective islands of sedges and grasses. Aquatic worms and invertebrates, without the freedom and protections afforded by water, have no choice but to make the best of a bad time by shifting into a holding state and waiting out the dry spell.

One can observe the effects of these adjustments, as if watching a card game being played. Nature controls the deck, and when conditions change the cards must be shuffled. The cards representing normal conditions must be placed on the bottom of the deck, and new cards brought into play to meet the changed conditions.

The lives of those marsh dwellers, the ones that are forced to continue sustaining life in the muck, become vulnerable to the stabbing beaks of predators. Killdeer, sandpipers, herons, and hungry egrets become sated. The smaller fishes, normally hidden in aquatic grasses and strands of algae, find themselves squeezed into an ever-decreasing volume of water, in exactly the same places where the larger, always-hungry fishes live.

Other than the constant chases and crowing of the egrets during this dire time of drought, the marsh is uncharacteristically quiet. It is silently coping, as if making every effort to expend as little energy as possible while stressed. It is an organism, after all. It will hold as many of its countless parts together as it can until rain comes and restorations begin.

AUGUST 21

The benthic muck of the Meadow shows more and more of its drying face every day, leaving little liquid to sparkle and wrinkle in answer to breezes.

Now, with little water cover, the wetland's frogs spend their time hunkered down within the many grass-covered oases. Late yesterday afternoon, a northern harrier came across the marsh, arriving like a giant butterfly, fluttering here, there, and over the ailing wetland. The harrier has come to seek out the newly exposed frogs, and the utter grace he displays while hunting is pure aerial ballet.

Moving from green island to green island, the hawk hovers scant feet above each potential habitat, peering downward into the enclosure, alert for any movement or recognition of prey. Swing left. Hover. Swing back right. Float in the air. All the time, his body feathers reflect shades of deep reds that shift into maroons and browns and rich grays. When he makes a turn, his tail spreads open like a Japanese fan.

AUGUST 22

The harrier performs again this morning. I watch him reduce the Meadow's population of beleaguered green frogs by four.

AUGUST 24

Time for snowshoes! Twice before, during conditions when wetlands have lost water cover, I've found that I can safely walk upon flats of mud with snowshoes to spread out my weight. Today, along with a University of Vermont intern similarly equipped, the drought-exposed face of Great Meadow was explored, and the day became one of shared wonder.

Two black bears, a bobcat, a coyote, a fox, deer, turtles, and muskrats had left their indentations on the wetland's parchment. Interpreting the "why" of tracks is a purely enjoyable exercise, doubly so when shared with a curious and intellectually gifted student.

Obviously the stressed condition of the wetland attracts wildlife, probably for reasons as simple as the fact that it's a new place, somewhere different to explore, a new habitat upon which senses may ignite anew. If food is found, that is a reward.

Furthermore, the baring of the marsh's droughty soil has resulted in an explosion of germinating seeds: grasses of multiple species, sedges, rushes, wild perennials, and shrubs, all displaying new stations for photosynthesizing. The drifts of arrow arum and aquatic knotweed blossoms are nothing short of stunning.

Then there are the ever-present, primary actors in this raw scene: the Meadow's birds. Swamp sparrows flit from buttonbush clump to leatherleaf patch. Tree swallows and barn swallows skim and bank, twist and sharp-turn toward, around, and just over our heads in their quests for flying bugs. While intent on watching a trio of killdeer stilt-walk across flats of mud, bill-jabbing repeatedly for who knows what edibles, an interruption snaps our heads to attention.

The killdeer threesome scatter and screech in alarm as a sharp-shinned hawk shoots out of the forest edge, making a beeline toward the inland plovers. The air around us seems to bristle with throes of fright and intent. Three, maybe four seconds at most, the killdeers escape with a preprogrammed scattering maneuver. Luckless, the cheated predator banks skyward, turning upward in wide circles.

Snowshoes squish in mud and muck, making hissing sounds when we step through the reaches of sedges and sharp-edged grasses. Frogs leap away and countless aquatic flies, erupt mist-like off the moist surfaces. Strange, when those blizzards of tiny flies return to the muddy muck, they instantly and completely disappear from sight, so perfect is their coloration.

When we approach beyond a watching distance, any great blue heron nearby will grunt and flap away to some point considered safer. Great egrets, perched atop stumps and not as skittish as the herons, watch us two-legged oddities shoe-paddle across their marshy world.

My life has been deeply enriched by mentoring many college interns through the years, most of them interested in environmental studies. It is more than a passing concern, however, that a preoccupation with electronics is closing the minds of younger generations, minds that should be opened *wide* for exploring what is offered by the *real* world. The real world desperately needs them.

When students look at a digital picture of a butterfly, as professionally composed as it might be, they are not seeing a *butterfly*. That picture cannot show that it is alive, vibrant, or has a life story that's remarkable beyond words. It is just a picture.

While sitting in a canoe with an intern, perhaps mid-marsh, at dawn, sounds riding on the morning mist offer lessons basic to that moment—but ones the intern is not yet aware of. The unseen teachers may be birds, may be insects, or the essence of the Meadow itself. It takes time to recognize nature's subtleties.

Both of us are here as students. The "Book of Great Meadow" contains an infinite number of chapters. Today's bright intern, who spends much of her spare time in natural settings, has benefitted beyond most her age from the teachings of her experiences. Her body holds several of the basic connections appearing in Great Meadow's book of stories. The old mentor may be a bit further along in that text—although with wisdom comes the understanding of how little he really knows.

I have changed many lives while out of doors, by simply requesting that my pupils stop: stop, and sit in a canoe, stand at an overlook, look into a spider's web, turn their ears towards some call. After many days, after many stops, after deeply probing conversations, their questions begin to spill out, stimulated by curiosity. The myriad voices of nature's teachers, once recognized, enter into minds eager to open wider than they ever have before.

SEPTEMBER 1

Mud-paddling on the marsh's muddy surface once again, letting the intern take the lead.

While walking upon countless organisms known to exist in the mud, I feel closer to Great Meadow's organic bulk on snowshoes than when I ice skate on its frozen surface. During this severe dryness, many of the internal workings of the Meadow have been put on hold.

My intern asks, "Does the marsh have an intelligence?" There is an example of a question born of curiosity. It is the right question. We take several minutes to analyze one another's thoughts.

During this morning's look 'round, we spy a three-year-old bald eagle perched atop an old pine snag. In its company, only a crooked branch away, was the first belted kingfisher of the year.

When we return to my house on this, the hottest day of the summer, what do we see fishing in the Meadow's brook channel? A common loon! Has Great Meadow ever accommodated this species?

We watch with the greatest interest, not wishing to miss a single moment of the unusual but worrisome visit. Cold, deep lakes are the usual habitat for loons. The conditions at Great Meadow are drought and really warm, shallow water. I wonder why this extraordinary and ancient bird has stopped here.

The loon also seems curious about his surroundings. At times, he inflates his swim bladder and rides high in the water, observing the 360° circle around his current position. Then he releases the air and sinks low in the water. Every few minutes, he dunks his head, scans underwater for prey, gives a powerful kick with his great feet, and dives. Two minutes, sometimes three pass before he surfaces, checks his surroundings, and dives again.

Later, during cocktail time on the deck, another act has diverted my attention. A sharp-shinned hawk, laboring on the wing while he carries a dead killdeer in his talons, chooses to perch atop an oak tree that grows directly over the deck. (Is this the same hawk that missed his killdeer lunch a few days ago?)

With my head crooked awkwardly back, binoculars vertical, I watch the hawk prepare his supper, step by step. As wing and tail and the largest of the killdeer's body feathers are plucked away, those same feathers drift down through the tree's branches and leaves, a few ending up right where Barbara and I sit. One secondary wing feather sticks in my martini! Another, a breast feather, lands in the cocktail dip. Others sprinkle the deck; an unusual form of precipitation.

The loon in the meantime has been absorbed into the darkness.

SEPTEMBER 2

No loon revealed in the dawn's light. He must be answering to the ancient insistence of migration, one peculiar to this species. Surprisingly, loons and hummingbirds share a lifestyle in that they fly their migratory routes alone.

I shiver when I realize that the loon that was swimming on Great Meadow yesterday will follow starlight, Earth's magnetism, coastlines, even scents from ocean currents and hints yet understood, to a place in the North Atlantic Ocean. In such a place, winter is harsh, but in that place the loon will find hundreds of other loons. The birds will form a sort of raft where, together, most will survive winter's demands. Where is yesterday's loon this morning?

SEPTEMBER 7

No rain. The sky is filled with dark air, yet the clouds, loaded as they are with tons of water, offer no relief to Great Meadow.

The Meadow's scenery, changing daily, remains pleasing to view—but many of its normal functions show severe stress from having been forced to shift into unnatural states. The killdeer and the ever-restless clusters of sandpipers have answered the migratory call of late summer and departed. Silence holds the Meadow in a firm grip both day and night, broken only occasionally by guttural squawks of the great blue herons. Barn and tree swallows still traverse the quiet air, seeking their insect prey.

From my Great Meadow perch, I try to think *into* and *onto* the troubled wetland: into the forced adjustments demanded of its habitats, and onto its multiple faces burned daily by exposure to the searing sunlight. My eyes scan, uneasily, the explosions of green vegetation that appear across the surface where water should be. I find I'm incapable of thinking deeper than what I can see, but I am sure that below the Meadow's surface, the drought is causing great stress in some species, while continuing to create opportunities for others. Knowing Great Meadow's surface story well, watching the current conditions come under stress troubles me. At the same time, I hope that the changes will bring moments new to my experience.

My brooding is sidetracked as a great egret cruises by, then another, followed by a third. It's as if they have been beckoned by something audible only to them. Such in-your-face beauty can't help but cause one's innards to sing. It upgrades my mood.

SEPTEMBER 18

Resident hummingbirds have left on their solo journeys.

SEPTEMBER 20

Spied two Diana butterflies on sedum growing in my garden today.

SEPTEMBER 22

Great Meadow: still in severe drought, still supplying sanctuary, still retaining its basic purpose, still a masterpiece—but in recent days mud cracks have disfigured its face.

SEPTEMBER 23

After a mountain hike (and a long nap on its summit) yesterday, my snowshoes call again.

My mud-oozing plod over Great Meadow takes place at midday under hot sun and in high-eighties temperatures. I don't expect much wildlife action.

Wrong. An osprey is hunting, and scores twice after three dives. The hunter pays me little attention, thereby affording me the sounds of both its splashdowns and the rustle of its feathers as it shakes off its load of water just overhead after becoming airborne.

Two eagles turn circles above the far end of the Meadow. After I shift my focus to watch the osprey for a moment, my eyes are unable to relocate the eagles. The reason: crowds of ducks keep flushing, negotiating wide turns around the wetland, practicing, strengthening, gearing up for the impending demands of migration.

From very fresh tracks across a mucky place, I note that wild things answered curiosity's call sometime during the night: A black she-bear with her cub paid a visit. The tracks clearly show their routes. The mother bear was exploring a new place, teaching. Her cub, following, stepped in her every track.

The bear tracks were crossed twice with those of a bobcat. Coyotes had also been called to the stressed wetland, though opportunity may have trumped curiosity. Here was mud art at its most expressive, demanding each work be studied and interpreted. Leaving those stories took willpower.

SEPTEMBER 28

While in bed, the time being two-ish, I heard moose grunting just a short distance down the steep hill towards the marsh. Leaf-shuffling and twig-snapping sounds confirmed my identification.

I went out on my deck to get closer to the goings-on, but the sound stopped momentarily and was replaced by odd sucking noises. The moose had left the forest edge and wandered out onto the marsh. Of course, it was hooves sinking into and being pulled out of the muck that was creating the vacuum music.

This morning, the tracks plainly show a family, probably bull, cow, and fawn. I think on it today; what a nice night for a family saunter through the mud.

OCTOBER 3

Upon returning from a train trip to Washington, D.C., I discover that a rainstorm has swirled its way up the eastern seaboard, gulped moisture in its travels, and unloaded a most welcome two or so inches of restorative wet onto Great Meadow's watershed.

The visual benefits to the Meadow have been immediate. Where a few days earlier my snowshoes pressed onto the Meadow's cracks of mud, water has now rejuvenated lives and provided opportunities long awaited by the multitudes whose very existence had been threatened by drought. The path back to normalcy has begun rapidly.

OCTOBER 4

The morning dawns brilliant. Water now fills hundreds of low places with thousands of puddles, each puddle collared by an almost instant greening of aquatic vegetation, so long in coming.

Five great egrets remain, and the harrier is still gliding, hunting for frogs. The great horned owl spoke from the white oak in the wee hours of the morning, and the eastern coyotes howled and yipped their notes, all in songs far too brief.

OCTOBER 5

From just before four o'clock in the morning, when Milky Way's suns outlined the constellation Big Dipper, to twelve hours later in the day as I stand looking over a newly painted Great Meadow, the light has been magical.

OCTOBER 7

The welcomed flush of rain has totally altered the visual appearance of the marsh. This morning, its liquid face is wind-crinkled just enough to reflect wiggly colors of shore-lined trees turned yellow.

Also in that picture are the white statues of egrets, the bottom-up tippings of Canadas, and the frequent flying flocks of orange-footed mallards. Yet even the superb, natural grace of the birds is relegated to understudy for this day's extraordinary sunset.

Ah, but what's that? Attention at Great Meadow can shift so quickly. There, just in front of me, seventeen jewels blinking on the water, seventeen sun-touched emerald feather patches flashing from a covey of green-winged teal.

These wonders of the duck clan are small, taking two to equal bulkier species like mallards and blacks. The green patch on their wings becomes iridescent when stroked by sunlight. Some report a green patch on the sides of their heads. Sunlight hasn't shown that to me yet.

OCTOBER 11

As noted before, duck hunters frequent Great Meadow now and then. Most often they go afield, hoping to return with dead birds and prove their prowess. Yet from what I have witnessed among most hunters here at Great Meadow, the equipment for their hunts tends to overshadow their experience and skill. I find that the waterfowl are little threatened.

There is entertainment value when the hunters show up, some with antics that are hard to watch with a straight face. Today's actor-gunner offers yet another example of human light-mindedness. The man must have read that aquatic fowl see very well, so he's camouflaged his entire canoe with a Spanish moss-type cloak. When he returns to his boat, he too is draped in "camo," cap to boot.

Well prepared for his quest, he now launches his floating island of disguise into the early dawn. Once at some chosen spot, his next move is to inflate Canada goose decoys; nine in number. Each plastic goose, with its own small anchor, is set afloat. The hunter then paddles his canoe to a clump of buttonbush, connecting a bungee cord to the stems to prevent any gust of wind from moving his station and revealing his position to his wary prey. With all plans and imposters in place, he pulls his moss-camo around his shoulders, his weapon readied. Undoubtedly, he's beginning to envision dinner, or at the very least, bragging rights.

There is, however, one slight problem. This hunter did not take field time to observe how Canada geese behave during gun season. Fall geese are extremely cautious—gunner-wise and gunner-wary. When landing on stopover waters, they stick close together, feeding only after either a long, watchful time, or after dark, and then only a few of any one flock can be seen feeding at one time. Sentinels are always chosen and on guard. (I expect few migrating geese really sleep.)

Instead of dropping his decoys in a tight group, this hunter spread his dozen bloated fakes over a large area. While watching, I see no geese settle onto Great Meadow—but if a skein had made a decision to drop down, those unnaturally dispersed decoys would certainly have been a warning, not an attraction. It appears it's the hunter who has been fooled.

OCTOBER 12

A spectacular fall day arrives with a nip: 28°F. The first white grass appears on the Meadow this morning. Two egrets still search for frogs.

OCTOBER 14

This morning, Great Meadow is empty of egrets. The much-needed rain has finally raised water levels in the wetland, providing greater cover for the egret's favored food. The lack of prey plus the occasional "BANG" from duck hunters combine with near-freezing temperatures to force those birds to seek warmer climes. I already miss their spots of white, scattered there, here, or over there. Their stay this year was an extended and beneficial one, brought on by drought and the availability of food. What games will next summer's climate change bring?

OCTOBER 26

I always savor my standard breakfast dish of fresh fruit as I sit next to the window that has the widest view of the Meadow. Early morning at this season of the year holds far less light than a month ago. The vast wetland, even though now nicely ponded, offers far fewer sightings of wild things, likely because their inner calendars dictate their presence elsewhere.

This morning, as my eyes strain to follow the wakes of three otters looping northward on some family mission, a black, rounded shape just outside and down from the window catches my eye. I cannot instantly recognize it in the gloom. At first glance, there's no movement, but a second, closer look defines the shape: Porcupine!

As I clarify the subject with my binoculars, I observe that the animal is holding something in his front paws, his head bent slightly forward. After holding that position for perhaps half a minute, the porcupine settles onto all four feet, brings his nose to the ground, waddles a few steps, stops, and then settles back onto his haunches once again, resuming his rounded shape.

Now, I can see what the animal is doing: Picking up, peeling, and eating the nutmeats of acorns. Surely not unusual, I expect. After all, porcupines shift diets many times during the three warmer seasons. Yet this is behavior I'd never seen, or even thought about. The porcupine continues his nut eating as squirrels do, but with sight variations. He picks the nut up with his teeth not his front paws, then, like a squirrel, settles back to peel the nut. He uses his front feet to manipulate the nut as he peels off the shell.

A porcupine's tail, however, is not flexible like a squirrel's, which has backward-slanting scales on the under-surface and plays the role of a stiff fifth foot—ideal for climbing and descending trees. Conversely, the stiff, lower vertebra in the porcupine's back forces him forward when in the sitting position; a bit awkward.

A later check of the peelings teaches me yet another lesson. The peeled chunks left by the porcupine are larger, by far, than those left by squirrels. However delectable this particular animal's night courses were, it's fun to think that this morning, he stopped for a four-nut dessert before retiring for the day.

OCTOBER 28

Busy land protection day, so immediately upon getting home to "the best spot in the world," and with the pesky clock racing towards late afternoon, I answer the beckoning of my new observation deck.

I call this addition my second office. I built it underneath the deck that overlooks the Meadow. It accommodates one chair, a low railing that doesn't interfere with my line of sight (but is just high enough so binoculars and journal can be kept handy), and when the clock permits, space for a late afternoon martini. For further comfort, a hammock is ready upon call. I will be spending a lot of time in this space in my dwindling days.

OCTOBER 29

With the temperature a balmy 70°F, October's brilliance splashes off the far hillside. The Meadow is tinted blue all the way to the far shore, and my mind and body are ready to sit. I open my journal and watch the morning's activities.

My eyes sweep across to the forest on the Meadow's far side and try to interpret the messages reflected there. The sunlit beauty of the oaks and beeches defies sufficient adjectives, so I close the journal and turn my thoughts to the two hundred or so Canada geese resting just across the Meadow, bathed in light.

Those geese spent the night flying through moonbeams, but this morning finds them little more than shadows moving through a thick mist that's spread itself over Great Meadow. Although I could hear goose gossip during breakfast, the opaque mist managed to hold the birds in near secrecy until the rays from the shooting sun revealed their sanctuary.

Completely exposed now, still the birds seem at ease. They preen, swim with abandon, pluck tidbits from underwater gardens, chat softly, and at times, engage in over-speak among themselves. The Meadow is their sanctuary on this morning, safe from gunners, perhaps a resting place to be stored in memories. I may well gaze upon these same birds next year.

NOVEMBER 11

The flock left just after four o'clock this morning, or so I assumed as I lay in bed and listened as their honkings faded away. I fell back asleep shortly afterwards—but when I got up a couple of hours later, the entire congregation was back on the Meadow!

For over two weeks, Great Meadow's welcoming habitat has held all those birds in its embrace. They gather each morning in the southern sections of the marsh, where they tip and refuel, separate into smaller clutches, and explore places that attract their curiosity. On occasion, wings are exercised, perhaps a half-hearted tip-up or two attempted, but generally they seem fancy-free and lazy.

Come midafternoon, some agreement passes among them to slowly come together. Watching all the birds filter back through the hundreds of bush-lined channels, answering to some high rendezvous call, is deeply engrossing; I am watching a staging of *purpose*.

Once the company forms, the birds take a leisurely paddle together to the northern portions of the Meadow. There, the next activity of this "purpose" is to schedule their afternoon forays.

This pattern, save for one day of blustery weather, has been followed for each of the fourteen days the geese have remained here. At least once each day, a fly-around of the perimeter of the Meadow is in order. At times, the entire flock takes to the air simultaneously. At other times, groups form and those parties fly helter-skelter, willy-nilly around the marsh like so many windblown leaves. Interflock conversation is always more raucous in those smaller flocks than when the entire flock flies as one. Oh, to have a little insight.

Last night was moonlit, with a steady west-northwest wind. As conditions became favorable, the majority ruled, and wings claimed the sky. The flock left for good near midnight. I'll be watching for their return next fall.

NOVEMBER 12

Wrong again! During a hike around the north end of the Meadow, I discover that the geese have returned once more from their night's foray. Assumptions are like thin ice: undependable.

NOVEMBER 13

An orchestra of goose voices interrupts my breakfast. From my overlook, this mass of flying geese can only be seen as reflected in Great Meadow's water. The compact skein is flying southward.

It is often difficult to see the actual flight of birds, even though you can clearly see their reflections in the water. If they are high enough aloft, they show well against the sky—but if their trajectory over the water is low, discerning them against the background of a mottled forest is not easy.

Many a time their reflection was all my eyes could find, never the birds themselves.

This morning's fly-past features an added component. Apparently, many of the geese hadn't evacuated their bowels before becoming airborne. Watching the water's surface dimple with scat brings smiles to my eyes. This time, watching reflections pays off.

NOVEMBER 14

A superb November day, and, once again, Great Meadow seems filled with geese. Same flock—right? Where do they party at night?

The Meadow's surface is glass-smooth, dimpled with geese again, but now, after a cursory count, I estimate that only half the original crowd has returned. The group stays put for the better part of an hour. They appear elegant as they float on Great Meadow's mirror. They are in constant, mumbling conversation. The banter is hardly audible. On occasion, I hear utterances of a bit higher register. Were I to choose a word to describe their musings, it would be "amicable." Simply said, a soft, peaceful listen.

NOVEMBER 23

The marsh is now frozen. The geese, still here this morning, are standing around on ice. The morning's sun shines at an angle that makes each bird appear to have a double image: bird and its long shadow. I get up to get a second cup of coffee and refill the sugar bowl—but during that short period of inattention, Great Meadow empties itself of its Canada geese for 2010. Another assumption?

DECEMBER 5

Red-bellied woodpecker: male, handsome. I didn't see that he had a lady this year.

DECEMBER 9

Fine snowfall: thirteen inches.

DECEMBER 11

I plod along this early morning on snowshoes. Pensive, my thoughts detach me from my usual rapt attention to anything a bit askew from the ordinary.

I am on my way to the old stump, the remains of the oft-mentioned Perch Tree, once Great Meadow's star attraction for its eagles, ospreys, and others. Judging from the diameter of that pine tree, its age upon death would have been about 125 years. When the beaver returned and flooded Great Meadow sometime during the 1950s that pine would have died due to the flood around its roots—but after dying, it had remained upright for over fifty years!

I honor the grand old tree for its utility. Sightings from it taught me countless lessons. It was the marsh's fulcrum for predatory birds. The loss of that tree had no bearing on Great Meadow's normal functions. Nevertheless, it had given me many moments worth thinking about.

My attention during my walk remains detached, noticing only now and then a track—bobcat, eastern coyote, then close to the shore, a hunting mink. It's there, against the bank of trees, that the sun's warmth gathers itself and gives me a hug. This bit of unexpected attention relieves me of my melancholy.

DECEMBER 16

Snow buntings! A first at Great Meadow that I've seen: a flock of plus or minus twenty, flitting in stops and skittering across the snow-covered ice. The last snow buntings I saw were in Svalbard, Norway.

DECEMBER 18

The red-bellied woodpecker readily eats sunflower seeds from the feeders on my deck. He'll swing onto a feeder, pick out a single seed, fly in a couple of swoops onto an oak limb (a different limb each time), and whack at the seed until it cracks. He is obviously inefficient at that task, even with his powerful beak. The tiny chickadees unhinge those same black sunflower seeds in a mere fraction of the time. The woodpecker should take notice.

Beaver lodge

2011

Snapping turtles hiss. Moose grunt.

FEBRUARY 6

And the 2011 winter has thus far performed as a proper New Hampshire winter should. Caution must be taken with the word "proper," as it is strictly dependent upon one's own assessment of winter.

For me, a proper winter is one with plenty of snow, accompanied by several days of below-zero temperatures. To date, 2011's winter has met those measures, with several storms in January and one significant snowfall in February, together covering the ground and everyone's roofs to worrisome depths—over forty inches. Snowshoes have been afforded their "proper" workouts, as the weather has remained cold and the snow pack satisfactorily soft for hiking.

Every morning, I continue to offer bread slathered with peanut butter (chunky), along with meat scraps and assorted table leftovers (considered safe) to *Corvids* and foxes. Both a gray and a red fox visit, but infrequently.

Overly cautious crows, high in the trees, bide their time for their turn at the snacks. While waiting, they usually watch as the local pair of ravens sail in and snatch away the largest morsels. When ravens fly around voicing their wide variety of opinions, I have no choice but to watch and to listen. I do not know I am about to witness raven hanky-panky, which amounts to a demonstration of "bait and switch."

A raven swoops in and snatches food in its beak, then sets its wings and coasts out over the snow-covered marsh. Upon landing, its entertaining strut-march begins. This usually includes the bird holding its head high while turning the food from side to side in its beak—telltale movements

2011'S WINTER BIRDS AT FEEDERS TO DATE:

6-8 bluejays
They have acted strangely all winter, inasmuch as they have lost their accustomed brazenness: This year they have shunned the feeders completely, instead feeding only sporadically under the deck, and even then acting skittish.
Brown creeper – seen on occasion but not on feeders
5-7 chickadees
40-60 goldfinches
40-60 pine siskins
3 white-breasted nuthatches
Downy woodpeckers – 1 male, 1 female
Hairy woodpeckers – 2 females, 1 male
Red-belliedwoodpecker–1 male
Of a scattering of neighbors who feed birds, four have reported red-bellies – all males.

signaling the beginning of a charade that's meant to mislead the crows that are watching.

* Misleading Act One: Show off food.
* Misleading Act Two: Delicately place morsel on the snow. (This act is often repeated several times.)
* Misleading Act Three: Strut around. Place food down. Use beak to gouge out a cavity in snow. Retrieve morsel of food that was placed on the snow. Place food in the hole.
* Misleading Act Four: Hold head and beak high with aplomb. Strut around hole in snow.
* Misleading Act Five: Dig another hole. Look around. Walk back to hole dug in Act Three. Remove food. Bring it to new hole. Drop it in (or, and you have to watch carefully, pretend to drop it in).
* Curtain Call: Bow. Zig-zag-walk away.

Charade over, the raven flies off! Has it flown away with the food, or did it leave the morsel in one of the holes? Surely, the crows pay close attention to the raven's hocus pocus, yet I've never seen a crow search for the hidden food.

FEBRUARY 24

Birds, birds, birds! While there is no shortage of species this winter, it is who is showing up that makes the specter of climate change continue to haunt my thoughts.

Yesterday, I purchased the third fifty-pound bag of thistle seed of this winter. At $58 per bag, that means I care a lot about hungry birds. Perhaps eighty to one hundred pine siskins and an equal number of goldfinches cloud my deck every day.

Two redpolls join the masses around nine o'clock.

Added a few more miles on snowshoes yesterday, taking in winter's perks.

MARCH 6

Two red-wings!

MARCH 13

Four common grackles.

MARCH 17

Three Canada geese voice their return as I am out feeding birds this morning.

Great Meadow is still thick with ice, topped with heavy snow, but I saw three otters yesterday fishing around an open slit in the ice. They were having success with their late afternoon fishing, with yellow perch on their menu.

MARCH 27

Barbara's birthday. The gift? A pair of bald eagles cruising over Great Meadow.

Their fly-by is accentuated by the arrival of two great blue herons. Since there is no open water, the herons drop onto the ice. I have never seen herons come in on ice. The climate is gumming up their timing. And the eagles apparently see no reason to stay.

MARCH 31

Ever since the much-mentioned Perch Tree was flattened by some bullying gust, it has been sorely missed. It was the major attraction across the entire, viewable southern half of Great Meadow, holding out limbs for many perching feet, especially those of ospreys and eagles.

This morning, while gingerly making my way over springy ice, I attempted to erect a substitute perch. The two perches I put up for smaller birds last year have been treated badly by ice. Now, each leans twenty degrees off center. However, I'm sure the birds have not noticed.

Allowing that the original Perch Tree was eighty feet tall with many large branches growing out of a two-foot-thick trunk, one glance at my skinny two-by-four affair, with its variety of thin limbs, automatically prompts the word "pitiful." I wonder, as I return to the house and look at the contraption from my observation deck, if it might just scare the birds away rather than attract them.

A pair of Canadas fly in at noon today and land on ice close to the shore. Might they be "our" nesting couple, returning once again? My heart hopes, as usual, that they are.

Late this afternoon, a single bald eagle (mature) moves around various tree perches, barely viewable from my windows.

After supper, I glance out across the Meadow. Such a warming sight: A pair of eagles, probably the ones seen a couple of days ago, perch close together on the top branch of a dead pine tree. Their distant silhouette resembles the shape of a heart. An odd picture: Two powerful predators huddled together, filling the lenses of my binoculars. I hand the instrument to Barbara and she is also touched by the late Valentine's Day image.

APRIL 4

Great Meadow's air has been turbulent, its waters dashing in rippling patterns all across its surface. Yesterday, I spotted the occasional duck, a couple of great blue herons, and a bald eagle picking the last remnants of winter's spoils out of the remaining ice.

At dawn today, Great Meadow's standard welcome is answered. My binoculars' field of view is filled with birds; an unexpected and welcome change. At least a dozen hooded mergansers present a mixture of sexes, all in costumes ready for spring's promenade. Mallards swim stiffly, half with green neckbands. Canadas, tall in their aloofness, number twenty.

A pair of wood ducks, tight together in the air, arc over the crowd of related swimmers and fly a sortie towards one of the nesting boxes. They perform their dual aerial inspection twice. On the third fly-by, the female breaks formation and flips into a box. Whatever adjustments or inspections she makes in the enclosure cause the structure to wiggle and shake.

I'm momentarily distracted by two otters who slide off an edge of ice and disappear, but are back on the ice shelf within seconds, making short work of brunch.

The duck box is now still. Is the duck inside? Or did she leave while my eyes were filled with otters?

A shadow alert! Just over the shore-side oaks coasts a fine, immature bald eagle, beautifully mottled with white feathers. It banks steeply and is soon on the ice exactly where the otters had just been enjoying brunch. Clearly, the eagle is taking advantage of the sloppy eating manners of the otters, finding bits and chunks of sushi scattered about.

Six great blue herons are stalking unwary fishes, but even their presence becomes secondary to a congress of crows having a chat-fest in a field at the northern end of the Meadow.

I take my winter feeding of birds seriously, doing those chores on behalf of their multitudes each morning. Those duties always include feeding a pair of ravens, along with the occasional crows. For the past week or so, fifty to seventy crows have come to take their leisure here at Great Meadow. They make their presence known with their raucous dialogue, and nothing escapes their attention.

When the feeding otters bring their catches up on the ice, a few crows fly to the spot and, like a gang of roustabouts up to no good, circle and harangue the otters with one barrage of caws after another throughout the entire meal. The otters pay no attention. The crows also play games with ducks, red-wings, and grackles. They apparently delight in interfering whenever and wherever they can. Respite generally comes in late afternoon, as crows tend to retire early, perhaps needing a good rest after a day's deviltry.

APRIL 9

Tree swallows and phoebes are back today. Again the swoop of an eagle catches my eye. An instant later, when the bird swings low to the water, a spray of water droplets flashes in the sunlight. It's not until I put my binoculars to my eyes that I see a real drama.

The action is on the far side of Great Meadow and the players are a pair of Canadas and an adult bald eagle. I've not seen an eagle attempt to prey on an adult goose before. As I watch, the eagle sweeps into a wide, banking turn, changing wing angles, swinging back again directly over the goose, circling, weighing possible approaches, seeming to gauge the proper angle for an attack.

Decision made: The eagle executes a sharp angle downward, sets its wings to glide low over the water, and heads directly toward the prey. But wait! The target goose, only a split second before the eagle's lethal talons strike, pushes itself up high in the water, neck stiffly extended to the max. This physical presentation obviously presents a formidable challenge, made directly in the face of the approaching predator! The eagle veers aside just before impact, pumps its great wings, and climbs steeply above the Meadow.

Still, the eagle is determined to make a meal of the goose. After several circling turns, it sets itself up for a second attack. On this attempt, it follows a slightly different method of approach. With wing flaps feathered

Pickerel weed on Great Meadow

and dropping quickly, this time it sets its run at the goose at a much lower angle over the water. Again, just before contact, the not-to-be-intimidated goose meets the foe head on, striking the water with its outstretched wings, projecting the entirety of its powerful body upward in direct confrontation before the adversary. Once again, the bird's startling act of defiance turns away the tormentor. That's grit.

The eagle, still majestic in every move, has not given up the hunt. It circles far against the marsh-side forest, swings in a wide arc south, and turns west to where it spots a second pair of geese.

These geese had apparently been watching, undoubtedly realizing that they too might become targets. The eagle heads for the new pair, dropping steeply and setting an arrow-like course toward them. As during the previous encounter, the new targets react at the last moment before impact—but their defensive tactic is slightly different. Both geese simultaneously stretch their necks to full length and rise up high in the water—but they also immediately extend their wings to the fullest. Again, the eagle veers off-target at the last second. To add to their formidable presentation, each goose literally screams as the eagle zooms past their heads. What is conveyed in their message is unknown, but the white-headed bird flaps northward, talons empty.

Canada geese invariably react to incidents, or encounters such as this, with a vigorous, standing-up wing flap. This triumph is no exception.

APRIL 16

Passing quickly by the breakfast window in a rapid flap, a red-tailed hawk shows that it has had a successful morning's hunt: a red squirrel dangles limp from its talons.

What's this? In close pursuit with equally emphatic wing strokes, another avian predator also has red squirrel breakfast in mind: a bald eagle, fully engaged in a traditional game of stealing food. Usually, ospreys are the subjects of eagles' aerial acts of theft—but not this morning. As the eagle closes the gap with the fleeing hawk, the chance of yet another successful food-heist becomes high. They disappear among tree branches in an instant. Any bets on who had breakfast?

Later this morning, eight of Great Meadow's wood ducks make their appearance high in an oak tree just off the deck. This mixed group moves from tree to tree, limb to limb. When landing, all strut forth and back on first one limb, then another, vigorously wagging their tails and stretching their necks repeatedly. The show becomes lost in the forest clutter of limbs, branches, and new leaves; spring fever perhaps?

APRIL 17

Three river otters are cruising and fishing, but mostly playing tag.

A chosen buttonbush in center-marsh has attracted a multitude of tree swallows. At least ten score wait for insects to warm up enough to fly and become bird food. While I never tire of being a part of such action, I do wonder about the circumstance that precipitates these scenes. The morning is very chilly (40°F), not good conditions for any winged insect species to be flying. Perhaps the birds simply gather on a convenient bush and chat awhile, while waiting for more activity, maybe finding warmth in numbers.

APRIL 21

Wind dominates Great Meadow's communities today, beginning at midday and steadily increasing to near gale force by late afternoon. There is no bird activity until close to four o'clock, when something noteworthy begins to unfold.

Whitecaps have been chasing one another across the entire breadth of the Meadow all day, each overblown foamy wave racing before its maker in strict compliance to the rate of force pushing it from behind. Toward evening, my eyes begin to see double white—white caps above white caps. The wind has taken charge of this scene, asking for attention.

Tree swallows! Hundreds of fluttering bodies flashing white, bodies slipping, sliding, dodging, dipping for insects, only inches above the dancing waves. Scanning the open waters with binoculars shows me dizzying multitudes of swallows. The water sparkles white, the swallows reflect white.

With the scene so surrealistic, my imagination adds distance to these images, placing me on a high prominence overlooking an expanse of ocean far below, its waves becoming giant breakers, the swallows easily envisioned as gulls.

APRIL 25

Early morning brings American toads' thrilling trilling, their voices numbering far more than memory remembers. Did last summer's extended dryness create a habitat alteration that greatly benefitted these toads?

Mergansers remain, but another oddity comes from the Canada geese who are whiling away their time here. On occasion, "our" local nesting pair flares up over trespassers who venture too near, crossing some imaginary line into their territory. That defense, however, is conditional. Many geese are hanging around in the vicinity, passing conversations forth and back. The nest guard must know each goose on a personal basis, as he is directly confrontational to some, chasing them far away, while he pays no attention to others. There is much to read into their relationships; selling them short is a mistake.

Today, a group of college students returns to join me in a woods look-around; it's an honor to be with these highly motivated young people. As is always the case, we find ample subject matter awaiting attention and curiosity. Our discoveries begin with an owl pellet, followed by three ant-cleaned skeletons of juvenile garter snakes, and later on, an army of springtails.

Over the past decade or so, few of the interns I have mentored come with readable signs that they understand the place that environment should hold in their lives. I'm more and more convinced that electronic "necessities" are replacing the wonders, needs, and teachings of living things.

The American toads are still in high voice at midday. Strange, they seldom call during daylight hours.

MAY 6

Varying weather, mostly cool, keeps the pastels of spring here at Great Meadow lovely, especially when days awake, and again when days retire.

"Our" pair of Canada geese continues to feed and pass time together within their prescribed space. No sign of nest preparation yet.

Toad calls are losing fervor, but a few solos still add sweetening to my bedtime.

MAY 7

Nine cormorants, doing coordinated fishing maneuvers, early in the morning.

Tomorrow, Barbara and I head for Belize and its bird-filled rain forests. . . .

MAY 25

. . .back to Great Meadow, and its continuing program of wonders.

As we left for Belize, the orchestra of American toads had been filling the eventide. Upon our return, it is the gray tree frogs voicing their paternal hopes. As with toads, the tree frogs also vastly outnumber any heard since my arrival here in the year 2000. Again, the question: Did last summer's drought bring about this phenomenon?

Something pleasing: While we were away, "our" pair of geese began nesting on last year's site. As of yesterday, the hen is setting and her faithful mate is patrolling.

Hummingbirds returned one of the days we were in Belize: two females, one male.

JUNE 11

Our Canada geese couple has hatched two goslings, leaving one unhatched egg on the nest. In direct contrast to the impressive numbers of amphibians voicing their spring pleasures, the hatch rate for Great Meadow's waterfowl is noticeably lower. I've observed:

Geese:

* Home pair: 2 goslings
* Two other pairs: one with 5, the other 3

Ducks:

* Wood ducks: 3 ducklings
* Hooded mergansers or mallards: none (as of yet)

This morning, the ravens brought their two mostly grown youngsters to feed on meat scraps. Those young made an unbelievable racket with their loud, rasping, begging calls. No complaints from me.

Late in the afternoon, an impressive large black bear is walking down my driveway, showing remarkable nonchalance. I watch as he saunters along, stopping after every few steps to turn his head to question why I am where I am. (I'm heading down our third-of-a-mile-long driveway to pick up the mail.)

As the bear ambles down the drive, I follow. When he stops and offers me an inquiring look, I stop. His easy saunter shows little if any concern, really not caring a whit about me.

After a couple hundred feet, the bear slides over a stonewall into a clutch of trees. I walk a bit faster and look into the woods where I last saw him; he's peering at me from behind a large pine tree. I continue along to the mailbox. I couldn't ask for better company.

JUNE 12

So many changes have come to Great Meadow due to last summer's dryness. New and vast stretches of aquatic grasses now cover areas where water should be. The local anglers, so successful at catching fish over the years, now find little action in response to the presentation of their fancy lures. I wonder, as I watch them casting, "Do they ever stop to listen to the music being offered from the amphibian orchestra as consolation?"

JUNE 15

Ravens again, but the two youngsters are now three. All five of this special *Corvus* crew raise a ruckus that slips through the house's walls and finds my ears. Those calls are impossible to ignore, but you can't just walk up to a window and watch. The keen eyesight of ravens usually catches even your most cautious attempt to peek. The call of the young birds is persistent, and although mostly full grown, each youngster begs attention from the adults with repeated wing fluttering, an act common among most young birds.

To have this raven family in such close proximity to my lookout windows both gives me pleasure and fascinates me. When they arrive at the feeding area, their impressive blackness, their formal posture, and their long wings, along with their propensity to hop from place to place, begs discreet attention. After reading many raven studies and knowing of their intellect and dedication to family (and to flock, when they're part of one), having this family close by and seeing them daily is one more gift. I'm off to get some road-kill to offer them, in thanks for their entertainment.

JUNE 21

At eight o'clock this morning the Earth tilted into summer. Five turtles—two painted, two snapping, and one Blanding's—are searching for suitable locations in which to deposit their eggs.

Once again, this particular visitor up from Great Meadow, the Blanding's turtle whose species is becoming so very scarce, has managed to return to the open spaces around my yard. I watch as she wanders across the unacceptable gravels of my driveway. She comes near to the flower bed where she once deposited eggs, stops, moves her head left, right, and down, but the scent does not satisfy. I feel empty as I watch.

While I observe her futile searches, two fine deer munch their way through the stunning blossoms of my favorite laurel bush. It seems they prefer pink to bright red, as the red laurel bush a yard away is ignored.

JULY 3

By the time spring tilts into summer and mating season dawns, Great Meadow has lost much of its cacophony of amphibious chatter. Its basin is full, with aquatic grasses and many species of floating pads creating an inviting habitat for the usual complement of birds.

But the question is, where are the birds?

* Geese: 5 families (and those have tiny broods)
* Black ducks: none seen
* Mallards: none seen
* Hooded mergansers: none seen

Only an occasional single duck swims on the Meadow. Silence in early summer foretells problems, problems not necessarily here at Great Meadow, but with the climate's anomalies and habitat loss elsewhere.

JULY 10

A pair of great egrets punctuate Great Meadow's sky like white exclamation marks. Their visit is unusually early.

JULY 22

The summer heat of 100°F and above that has plagued the southwest has now climbed north over the continent's parallels into New England. The embrace of heat arrived yesterday and is creating sweats today, with more promised for tomorrow.

Early this morning, five Canadas were out among the grassy islands, tipping for early breakfast before retreating into the comfort of shady reeds.

For the second morning, tree swallows are congregating. They are in high voice, with this year's hatchlings learning about flock etiquette and possibly about the excitement of the migration that is ahead of them. Great Meadow's insect nutrients will provide their share of the endurance protein needed by the swallows for the journey to the Carolinas—or with the temperatures changing will they perhaps winter farther north in Virginia?

AUGUST 13

With a cool front defeating the hot spell, this morning has a September flavor.

AUGUST 16

The "house" hummingbirds have brought forth two offspring that all day, every day try to emulate the actions of their scatter-brained parents. The deck, once again, has become a place hardly conducive to relaxation.

From my superb under-deck hideaway, I challenge myself to keep my eyes on a single tree swallow as it moves over the Meadow. With so many birds zigging and zagging, I give up quickly.

Instead, I watch an activity I can follow: A large snapping turtle lum-

Snapping Turtle - Eons of relatives

Snapping Turtle egg

bers up onto a mud-bound log and collapses. It's there for a sunbath. Three great blues, each especially vocal this morning, now distract me from turtle-watching as they fly with their usual resolution from one station of promise to the next.

Despite the morning's brightness, something here at Great Meadow is unsettled. The stage continues to lack the usual complement of actors. The squeaks of the hummers, the squawks bleated from the herons, both fade quickly into the far-too-silent Meadow's morning.

August continues lonely. The broad pastures with their ripened seeds, with their tender rootlets all populated with insects in various stages of tastiness; the marsh with its places to swim, with its nooks in which to hide and sleep, with grassy hummocks offering safe niches on which to rest and watch—all are going unused.

True, the swallows' movements are eye-catching, and herons and buzzing hummers ask for attention, but there should be much more. It seems that those "many mores" occurred yesterday, and like all the yesterdays that once were so full of Meadow-connected lives, I ask, "Might those days never return?" "Will only memories be left?" It seems that the Meadow's vibration has gone mostly silent, its vitality tempered.

I've not seen a duck for days.

AUGUST 25

The late afternoon sky, mottled with streaks of gray, serves as a backdrop for a few remaining tree swallows. One wood duck squeaks at fly-off. One bullfrog grinds. Those sparse whispers only reinforce who wasn't at Great Meadow over this silent season.

Today, waiting is in the air. A hurricane is predicted to blow through and douse this region. I think about the tree I call Old White Oak—not a leaf in her crown quivers. She has withstood a wide cast of winds for at least a century and a half, probably longer. She is an elder here at Great Meadow, ancient, strong, graceful, along with deep integrity of form and function. Pendulous knobs bulging breast-like from mid-trunk reinforce her attractiveness, reminding thoughtful viewers of the tree's innate ability to repair serious trunk wounds where limbs were ripped away by the force of past storms.

This storm, Irene by name, turns her wrath westward, testing the ancient tree with her bluster. I watch the oak's old, stiff branches sway and creak. I listen to the whistles of her leaves as they sing bravely in the face of Irene's best. Irene moves on. Old White Oak adds one more storm to the latest ring marked in the cross-section of her enormous trunk.

AUGUST 28

Kestrels! By every report, populations of these often-called sparrow hawks have plummeted, so to have five of these painted predators spending time in trees just overhead, and to watch them play aerial tag over Great Meadow's precious spaces, is thrilling.

I expect the five are of one family. They are very vocal as they flit though the treetops. Like the raucous parrots in Belize and in the Amazon, their conversations are raspy and squawky. With the many birds I've seen around the world, it seems that the more colorful they are, the more their voices lack tones pleasing to the human ear.

American bitterns appear in midafternoon—three birds hunting, piercing their separate ways through the Meadow's thick grasses. I haven't glimpsed the species all summer, so these three herons must be from elsewhere, now moving, now answering to the urging of the season.

Later a pied-billed grebe shows up, always a treat to see. I expect this bird is complying with the same impulse as the bitterns.

SEPTEMBER 30

We often hear a porcupine outside, right under our bedroom window, just as dawn is lighting the day. It *hums* its own soft notes as it pokes along! The hums do not vary in pitch. They are emitted in a series of four or five as the animal trudges back to its daytime den after a long evening of foraging. I've watched on three mornings, in pleasant amazement, as the prickly vocalizer passed under our window humming. Damn special.

My porcupine is not alone here at Great Meadow with its dulcet humming at dawn. On occasion, my old ears are alerted to the calling of a long-eared owl. Yesterday, well before dawn took charge, the male owl was calling, calling, in notes so feeble they barely rippled through the faint light of dawn to my ears. I strained to hear if some response answered his request. None did.

Hearing the voices of the owl and the porcupine during the same predawn would give any day a positive start.

OCTOBER 9

My ears are barely awake enough to, once again, catch the gentle calls of Great Meadow's long-eared owl. I slide out of bed, stand at the open door to the screened porch, and hope. My hope is answered, as a call creeps in through the morning's moist air from the male. And then it happens—an answer from a female, or at least I hear something that sounds like an answer.

The male calls again. Yes! Yes! To my left, somewhere in a tree along the shore, comes the same slight response. One note. One oh-so-brief note. One.

The male sends his barely audible call once more. The female's reply? One soft, soft "uh."

I've experienced nothing in my years of listening that has been so touching as those voices. Simply poignant. I go back to bed, put my hands behind my head, and let those generously shared moments from nature sink in.

OCTOBER 14

As the leaf-flaming forest around Great Meadow morphs into the spectacular, Barbara and I are awakened once more by "blam, blam, blam" explosions caused by the trigger fingers of duck hunters. If all were true sportsmen in quest of waterfowl for their tables, I would express no objection. However, more often than not, the gunners I've watched on Great Meadow simply come to kill.

This morning offers no exception. Three Canadas are gunned down, one dying instantly and two dropping into the water with injured wings in response to a fusillade. The two birds do all they can to escape by deflating their air sacs, even submerging completely on two occasions. Since that escape tactic can only carry the injured birds a short distance under water, the hunters, in their engine-powered boat, take up the chase.

Instead of killing the birds when they surface for air, the hunters discover great fun in chasing the doomed geese. After the birds become exhausted, the hunters fire at close range, scattering two heads and two necks across the surface of the marsh. They leave the bodies behind.

OCTOBER 15

The same hunters returned this morning, pre-dawn, but as they motored their boat up-marsh, the Canadas who spent the night on the Meadow took flight well beyond shooting range.

OCTOBER 25

The resident geese are back on schedule. On mornings not disturbed by hunters, the current Great Meadow flock of thirty birds has quickly adjusted to after-hunting-season days. The Meadow's geese answer unwittingly to Earth's time. The internal clocks of the birds are calibrated every day, and they have no option but to act upon their migratory instincts. If I pay close attention, unless the weather grounds the geese or hunters are lurking about, I can predict almost to the minute when the "time to fly" message will stir throughout the flock.

Each morning, the geese voice their intentions to move shortly before taking flight. The next morning, vocalizing time is advanced by roughly-five minutes. Sometimes, all thirty birds burst into flight—but the routine of leaving Great Meadow is never the same.

Even if the entire flock takes to the air together, that does not mean that every bird has the same destination in mind. Some quickly break away from the formation and head onward; many times one, two, or more birds stay on the water.

When Great Meadow became part of my life, I was thrilled to simply watch as its birds, species by species, added chapters to my story. After years of refining my observations, it is unquestionably clear that geese are not automatons, not even close. They display individual traits; they relate to one another and to family both on the ground and when their skein is in flight. They have taught me not to question whether or not they possess wisdom, nor whether or not they can play games—although I'm hard put not to think that much of the latter seem to have some less than honorable intent.

OCTOBER 30

Again, I repeat to myself: With the subject of weather, long-time averages no longer apply at Great Meadow. Last night, a storm crept in, working during the dark night to surprise most New Hampshire folks with twenty-three inches of snow—a white Halloween trick. While the Meadow's living world is inconvenienced not a whit, humans across the morning's whitened world scramble to make do without electricity, all the time putting up with disappointed kids whose Halloween costumes have to go back in the closet, at least for now.

While the shallow acres of Great Meadow are laden with snow-slush, its open water sections are stirred by the webbed feet of a dozen Canada geese; their resilience nice to see.

NOVEMBER 22

Thanksgiving arrives, waistlines expand, and along with the feast comes weather far too warm for the closing of November. Warm weather again, this late in the season, presents winter preparation problems for many animal and plant species.

Great Meadow's Canada geese continue to paddle and converse in now slush-free water. Hunters come too—but, as if clairvoyant, the birds melt away into the greenery someplace, giving the hunters, patiently waiting in their camouflaged suits and boats, nothing to do but smoke and, on occasion, pee into the Meadow. When the hunters leave, the geese show themselves and continue with their own Thanksgiving smorgasbord.

With temperatures being un-November-like, the facts of climate change and tomorrow's uncertainties do not cease to prey on my mind. Now in late fall, Great Meadow is like a huge blossom preparing for spring. Its vital reproductive parts are working. Its mission to provide the reward of sustenance next spring is already underway.

Question: When Great Meadow brings forth her multitudes of nutrient-bearing flora and fauna in the coming spring, will a full complement of hungry species come to take advantage of the bounty? An unlimited supply of available nutrients waited for consumers last spring. Few came.

DECEMBER 12

Tonight, a full moon sends its gentle light in a band across Great Meadow's calm water.

Nice, but wrong. It is mid-December. Moonlight should be skating on ice, not surfing on water. Outside temperatures remain well above freezing. Old averages continue to lose their places in statistics.

Muskrats are still lodge-building their winter hangout. Beavers still V through the pond, headed for wherever. During the three previous days, two bald eagles, one immature, the other a fine, impressive adult, swept their shadows across Great Meadow's liquid surface. They've since flown away to another address.

A male red-bellied woodpecker, absent all summer and fall until last week, is making daily appearances at the banquet I provide on my deck. That bird, with his roseate cap and lavishly striped sweep of black and white feathers across his entire back, gives a touch of formality to the flighty crowd of smaller birds whose dining manners amount to little more than grab-and-dash.

South end beaver lodge

2012

Wild turkeys gobble. Hooded mergansers mutter.

JANUARY 6

Finally, a gift from a reluctant winter—ice. I eagerly accept and go out to explore. I find surprising changes. During the long, unseasonably warm days in December, Great Meadow's two families of beavers must have been very busy, as I can see that major construction has already been completed. The south-end beavers have abandoned their old quarters and formed an entirely new lodge in a new location.

The old lodge, in its last stages of livability due to rotting and general decomposition, is now serving as habitat for otters. Two dens have been dug into the above-water sides, and they might serve as temporary digs every time the otter family passes through on their territorial hunting route.

The north-end beavers haven't built a new structure—they've refurbished an old one located close the Meadow's outlet dam. Perhaps it was their parents' home, their birthplace.

JANUARY 8

No snow; temperatures in the high forties. The coldest it's been to date this winter: 24°F. Ice has formed, but I question its invitation.

With no cover of snow in the forest, ample food is available for birds. Those that do come to my feeding stations visit only half-heartedly, sort of like folks going to the local social club just to be with some company. Oftentimes, the birds just perch on one feeder or another, not bothering with the seeds at all. Time to gossip, time to preen . . . time to pose for the folks sitting at the breakfast table who spend good money to feed them.

Red-bellied woodpecker; female. May she find the resident male up to her standards.

Raven feeding has begun, but they come only sporadically, flying by and immediately circling back to double-check for any food that's been offered. Thus far, only once has a bird dropped down and flown off with a snack.

JANUARY 13

A dreary afternoon following a morning of thick, heavy gray air. Yesterday's storm contribution of six inches of gooey snow left latent clouds over the land in formations tight enough to shorten the afternoon. Those same stubborn clouds also refused the usual brightness of the morning's light. Gray continues to rule the day, sulking and spitting rain, and pelting the Meadow with both sleet and hail.

Like an invisible conductor, the north wind apparently tires of the tune and raises its baton. The forest responds by shaking branches free of snow, each limb rebounding with satisfaction and relief. As the released snow tumbles earthward in clumps, the north wind adds a gust of power to each falling mass, shattering any semblance of form and instantly transforming the falling snow into horizontal mini-blizzards.

A clearing to the west of Great Meadow opens, giving a beam of sunlight the opportunity to shine upon the grayness, withering it all in a matter of minutes.

The now snow-free conifers along the far shore appear to ignite. Ribbons of windblown snow, some running the entire length of the Meadow, are illuminated, their whiteness whitened. Above the horizon, great cumulus giants wearing curly crowns of pink, pull along their sagging, dirigible-like gray bases, towing them out of sight.

Four beavers come into view in the marsh's now unfiltered sunlight. With the continued moderate weather, the animals have been able to maintain an ice-free swim-way along the far shore, giving them access to juicy rhizomes and grass rootlets that grow in the muck of the marsh.

JANUARY 21

Three days ago, the meager cover of snow on the marsh was rained on. That moisture then froze, creating a crinkly crust. My boots broke through that cover, hitting solid ice just below and making for a couple of days of grumbling during my walkabouts.

Today, to make my meadow hike less strenuous, I lace on a pair of small, cleated snowshoes. With that advantage, I am able to cruise easily on top of the flimsy crust, helping me to forget yesterday's dour mood.

Standing on the frozen crust of the Meadow, I spend some minutes watching the activities of a flock of goldfinches. While at the feeders, they feed ravenously, but only in thirty-second spurts before exploding into the sky, often returning to the very branch they were perched on moments before as if hitched together by invisible bungee cords. Usually, the flock twitters and fidgets a few seconds before looping back to feed. Some days, they head out far over Great Meadow, beyond sight.

Today, I find myself almost directly underneath those clustered dots of yellow, a privileged observer of a goldfinch clan exercising some secret ritual over the Meadow. Their loops and swoops, sometimes tight, sometimes stretched far apart, play out just above me. Even their twittering, each call usually sent at the low point in their flight loop, becomes an unbroken sound, close to a whistle.

I'm spellbound for over a minute as I watch the bright finches burn the very calories I spend good money providing. Could it be that the show I've just witnessed is their way of giving thanks?

JANUARY 24

At 45°F the frosting of snow is, again, melting. Winter has yet to visit Great Meadow with any intent to stay.

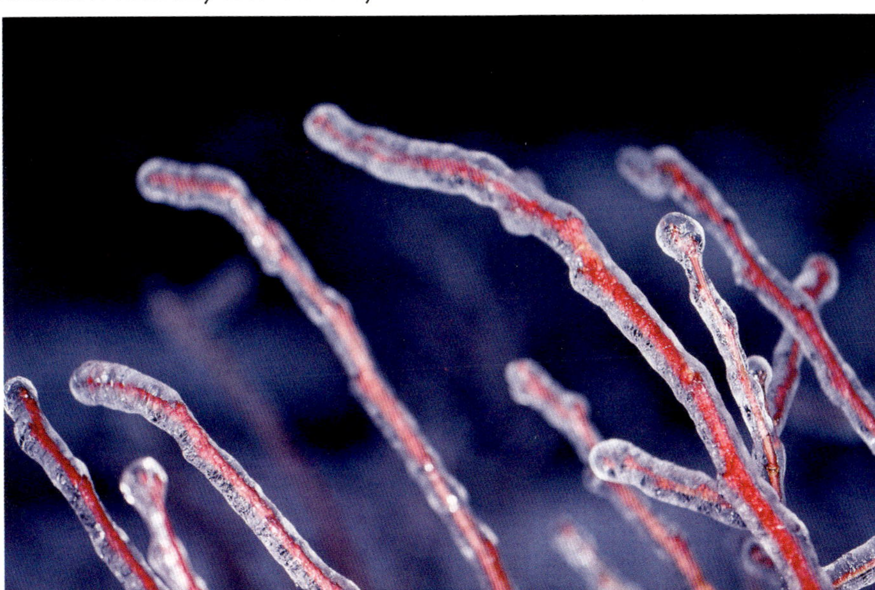

Frozen blueberry twigs

JANUARY 26

A dreary morning, 40°F, but the day quickly improves as Great Meadow calls and I answer.

With the marsh now covered in slick ice, I dig out my ice grippers. (I seldom resort to this equipment for the simple reason that they make a lot of unwanted noise.) I clank along across the ice southward, acknowledge the stump of the long-gone Perch Tree, and then continue into the far cove. Remnants of crusty snow hug the shaded edges of the Meadow, revealing the route taken by a mink, the tracks crossed by an eastern coyote.

As I turn to head east along the southern shore, I spot two beavers eating on the ice-edge near their lodge, located a couple hundred yards away. The beavers twirl small branches in their paws as their teeth sheer away the tasty bark. I move very slowly towards them, using shore-edge blueberry bushes as cover. I get within one hundred feet and settle down on a convenient rock to enjoy the unscheduled entertainment.

The beavers remain intent upon their feeding. One, then the other, slides off the ice, disappears, and then reappears with a new stick snack that must have been stashed away around the base of their lodge a couple months ago.

The half hour that passes would seem reward enough for one day, but there is more. When both beavers duck underwater, I scratch along the ice and get closer to their lodge, which is surrounded by a reach of open water. (When winter allows, beavers work to keep an opening around their lodges ice-free. That gives them freedoms not possible when their lodge is ice-bound.)

While I stand motionless, both beavers appear at the edge of the ice shelf and eye my dark form. I call up an old chant, one I've used over many years when I've encountered wildlife. By refining the monotone of the hummed chant, something with the sound gives pause, or as is often the case, actually elicits curiosity from the animal in sight. Deer often will become curious enough to move a few steps closer, providing my scent keeps to itself. The chant only works if I see the animal before I'm seen.

Here, back a distance from the edge of ice, I begin my chant. Less than half a minute passes before both beavers surface. I continue. They swim forth and back very slowly, and on two occasions stop next to the ice, as if listening. I chant on for several minutes. Eventually, they reach the limits of their curiosity, slide under water, and are not seen again.

I continue along the edge of open water where I saw the beavers feeding. There on the ice are pieces of lily pad rootstock, all the fleshy parts eaten.

They had chosen roots with diameters of about an inch. Those, I expect, were young and tender, and contained the minerals that beaver relish. Why wouldn't such a salad be tasty?

From there, I gingerly walk deep into the far cove and along the Meadow's shore, which abuts a pasture. Otter were there recently, made obvious by the shiny, silver fish scales scattered around and on several of their feeding stations. These are the good things of the day.

The bad thing happens as I am headed back across the Meadow for home. Unaware that the ice cleat on my left boot dropped off mid-step, my boot contacts the ice and I'm down in an instant. Out flies my left arm to protect my fall—but the slick ice shoots my hand forward, thus extending my shoulder joint well beyond where the bones are able to go. The shoulder socket separates, leaving me in yelping pain.

Thoughts get jumbled when the brain is busy coping with physical shock. I am in view of my home a third of a mile away. Thoughts race: I hope Barbara did not look out and see me on the ice. . . I don't have my cell phone. . .if I *had* my cell phone, she would not handle the situation calmly. . . I've probably messed up our trip to Chile coming up in two weeks.

I'm down. Getting up on slick ice, one-armed, is going to be a problem. I have to jerk myself along on the ice to a distant clump of buttonbush that has strong stiff stalks I can hold onto with my good hand. Once there, I manage to right myself.

Now what? One look across the one-third of a mile of ice between me and home and I know that getting there on one cleat is not a possibility. Putting the separated cleat back on is also out of the question. The only option is to make my way off the ice, put one foot ahead of the other and climb up two steep pastures to a neighbor's house.

That walk is a struggle. The slightest movement of my sorry arm somehow weakens my legs. An hour later I rap my walking stick on the neighbor's door; dear old Alice is home. From there I call 9-1-1, and a twenty-five-mile ambulance trip to an emergency room to get fixed follows.

FEBRUARY 6

I've written few words in this Great Meadow journal on personal matters; its purpose has been to share what a body endowed with sensitivities towards nature can clearly see. However, after my tumble, I'm stuck inside to interpret what is happening out-of-doors. My arm remains of little use, especially the hand and fingers.

Not being able to stand on the Meadow's ice, to have the sense of being one with that remarkable whole, has led to darker thoughts about my own longevity. At a ripening eighty-two, I realize each hour spent outside is an irreplaceable gift. I fully expect my arm to improve incrementally, but having heard the docs mutter something like "gaining some use may take a couple to three years" nags at my thoughts.

During my unsettling convalescence, I walk the rural road that my driveway joined in 2000. Despite the chance of a tumble, the thick forest that embraces the dirt road extends a welcome I cannot ignore. I take due caution, glad to get into the woods where I can feel the natural environment's comfort, its voices, its healing welcome.

FEBRUARY 20

So much is not as it should be with the climate. No snow, and day follows day here in mid-February that really belong in early April. My daily, cautionary hikes are on bare ground, with temperatures well above freezing. With the weather discombobulated, and a body part the same, that's two that need fixing.

FEBRUARY 22

It is good that the days are dawning earlier because Great Meadow's stage becomes visible during breakfast. This morning, a happening of note accompanies my eggs dropped on toast.

Appearing from behind the near-forested edge of hemlock branches, a line of wild turkeys make their way across the Meadow's slippery ice. One by one following their lady leader, each bird dutifully aligned in formation, come seventeen members of the big bird platoon. Such a parade puts a smiley edge on an otherwise drab, dreary morning.

As the turkey procession stretches out across the gray ice, the lead female, with measured steps, makes straight for the far shore, maybe a tenth of a mile distant—but progress is tedious. Here is a performance based on the finest traditions of slapstick comedy, with clowns disguised as turkeys. Like a conga line of drunks, the actors' feet keep slipping, tipping the turkeys down repeatedly: Slip to the right, flail the right wing; slip to the left, flail the left wing. The act is top-notch silent humor. Of course it probably does not occur to the leader of this line of comics to set an example for her company: *Fly!*

The procession finally comes to open water along the forested shore. There, the turkeys begin to bunch up, standing around on the perimeter of ice. A couple of decisions have to be made: "How best to cross the expanse of water?" and "Who will lead the way?"

As I watch through my binoculars, the turkey that leads the retinue across the marsh is the first to answer both questions. She seems to have suddenly remembered her wings! Her decisive action is example enough, and the rest follow suit, lurching into the air. Is the show over? Not quite.

A quarter mile behind, also having difficulties keeping their footing, come three gobblers. Their advance is a stop-and-look-around affair. As with male turkeys that have ladies in mind, they get lost in their own self-importance. Strutting while formally presenting an inflated body of feathers is hard to pull off when every step is on slick ice. The attempted pomp of this trio of suitors becomes little more than a scramble of feathers—but strut and slip they do, making their way towards the troupe of ladies in waiting.

But those ladies are now off in the forest. When the gobblers finally notice that the loose assemblage of girls has disappeared over a tree-lined ridge, their demeanor instantly changes, purpose and dignity restored. All three remember *their* wings, and use them.

Great breakfast entertainment.

FEBRUARY 23

Three Canada geese; they have returned far off schedule.

FEBRUARY 24

Off to Chile, Argentina, and Uruguay. . . .

MARCH 12

. . . Twelve inches of snow are reported to have fallen while we were away, but none remained to greet us at Great Meadow.

MARCH 13

With understandable fatigue, I manage a short walk along the Meadow. Shockingly, it reveals yet another anomaly of climate and how it confuses wildlife, especially those that migrate.

I hear the "crick, crick" calls of common grackles. At no time in years remembered have grackles preceded red-wings to our wetland haunts here in New Hampshire. Their internal programs are out of sync.

MARCH 14

Two days ago, the shadows of eight herring gulls reflected on the marsh; too early. When these first eight arrived, they simply sat together on a small patch of remaining ice. Yesterday there were twenty-four, and now today, three or so score of new arrivals are doing the same—between fifty and sixty in total. Of course, reasons and communication pertaining to some attraction or benefits here at Great Meadow have led them here, but *never* in these numbers and *never* this early in the year. The "nevers" keep adding up.

As I write in my journal this afternoon, the only remaining ice raft is barely wide enough to hold all the gulls at one time. There is much jostling for space and each bird seems to be upset with his neighbor's pushiness. Some message travels throughout the flock and the entire throng lifts off in a volley of wings.

Up they rise, white wings layered against forest green and sky blue. The flight pattern they display reminds me of a square dance, where the dancers keep changing partners as they rotate around the dance floor. As the graceful gulls intermingle, they do so effortlessly, silently, serenely, all the while playing in a slight breeze that slips across Great Meadow's dance space.

About two minutes into their do-si-dos, the flock complies with an order to return to the ice. Like parachutists heading for a drop zone, the gulls cup their wings and touch down once more on the same unaccommodating patch of ice they had just left. The fussing and jockeying for space start again.

MARCH 15

Ah, finally, early morning has brought the arrival of fourteen male hooded mergansers.

How very proper the males are, strutting their high-crested, fancy costumed bodies before a half-dozen ladies-in-waiting. The gentlemen dandies swim close together, performing like a company of dancers. They seem to mirror one another's moves, stretching their startling white head fans to the limit, emphasizing the effect with rapid head bobbing.

Water dipping is next, done while moving as close to the females as possible, spraying them a bit in passing. Spins are enacted, stirring the water enough to bounce the females a tad, and those fancy spins are often followed by emphatic body dunks.

Short-speed swims covering no more than a yard are also included in the show; however, like a rubber band stretched to its limit, the dash away has a predetermined length. When the bird reaches that limit, he instantly turns around and speed-splashes back to the female in a climactic end to the effort. Despite all the grandstanding, not one lady has shown the slightest inclination of interest—but perhaps that's only meant to encourage greater zeal in the would-be suitors. In much of the bird world, females tend to be purposefully distant.

TODAY'S BIRDS:

3 common mergansers
1 pair of mallards
1 wood duck
1 great blue heron

MARCH 16

Time to attend to the three duck boxes. Normally, such preparations aren't in order until April 1, but the record-setting heat over the past two weeks (high sixties to low seventies) has to upset mating schedules and internal physiologies.

MARCH 18

Surprise! I get out a stepladder, peek into a nesting box, and see a completed nest! That can't be. She must have built it in the dead of night, as I've not caught a single glimpse of her comings or goings. Of course, I examine it: masses of leaf bits (not one whole leaf), a variety of twigs and woven grasses, and bits of down (her own?)—all close to a month early.

MARCH 19

Yesterday afternoon and last evening: spring peepers. Far too early, but then again, yesterday's temperature reached 70°F.

For the calling frogs, there is but one purpose to their messages—attract mates—but now, for this listener, the usual comfort I feel from these early calls is tainted with uneasiness. Both the outer and inner natural forces of Great Meadow react to climate when it changes. This year, peepers have been forced to adjust to an early wake-up call yet again. Are their genes adjusting?

MARCH 20

Spring brings back "our" phoebe.

MARCH 21

This afternoon is a difficult time for "our" resident pair of nesting Canada geese. They spend midday together, with the gander staying close to the nest. The scene changes when two male geese, who seem to have some shared purpose, approach the mated pair. The hen gets off her nest and joins her male partner—but the intruders bully their way close to the female while her mate displays considerable agitation.

With the female clearly the object of the ruffians' attention, loud honking erupts, with threatening postures and various forms of neck twisting exhibited by all parties. That engagement is followed up with rapid beak waggling and tongue waving, with no participant seeming to back down.

The posturing goes on for several minutes, until the action abruptly changes: The defending male suddenly lifts off into the air and rams his adversaries. When the intruders retreat only a short distance, the male, now on offense, forces them into flight once again. It takes three such flourishes to send the gang of two off across the marsh.

Satisfied, the male, now flush with his job well done, flies back to his lady—a chorus of honks and displays of wing-flapping having decisively proven his worthiness, yet again.

MARCH 27

Much cooler: forties.

Tree swallows.

MARCH 28

Common mergansers: two males, six females.

Red-bellied woodpecker today: male.

"Normal" late-March weather has arrived: cloudy, showers often mixed with the standard offerings of sleet, freezing rain, and hail. Meantime, Great Meadow's now ice-free surface is disturbed little by paddling birds. Eerily quiet in this usually critical and busy season.

MARCH 31

Weather staying seasonal. With this upgrade, migrant bird numbers have marginally increased over the Meadow.

The "home" pair of Canada geese pays rapt attention to their old nest site this afternoon. The female is walking upon the nest as the male seems to be assessing the sanctity of the territory beyond the site.

It's difficult to avoid keeping my binoculars off a great blue heron that's standing close by. This bird sports feathered finery that flows in long, graceful curves down its neck and across its flanks. These birds are an extraordinary species, inasmuch as they can show a gangly awkwardness during initial takeoff, but once underway, their command of slow grace in flight is moving.

A bevy of herring gulls returns to sprinkle Great Meadow with their fits of floating and their starts of syncopated flight.

APRIL 1

The mother goose is down to business on her nest. The current weather may be good for a nesting goose today, but unless rain comes soon, the nest will cease being separated from the land by water, providing easy access to predators like raccoons, foxes, and others.

APRIL 14

Special discovery: a hooded merganser male is alone on the water, which is a station he regularly takes when his mate is in a nesting box. I watch his actions for about a quarter of an hour and hear a call, but do not recognize it. I am more than a little surprised when I discover the source: The male merganser is calling!

His call is distinctly frog-like, really not too different from the grinding utterances emitted by pickerel frogs. This bird's "grind" is preceded by a "tuk, tuk" call emitted through an open beak. Immediately following those staccato notes, the merganser tosses his head back, lifts his body up, spreads his grand hood and sends forth this two- to three-second wheezy call. The pitch is higher than that of the pickerel frog and shorter in duration, but definitely similar.

Do hooded mergansers have a "quack?" Is the call I witness strictly one of assurance to his mate, who surely understands?

APRIL 15

First dragonfly hatched: common green darner.

Osprey carrying a large, white-bellied hornpout over Great Meadow.

MAY 2

Statistics read that the incubation period for Canada geese is twenty-four to twenty-eight days. Our female went on her nest March 31. Yesterday, her stalwart mate perched next to the nest site and remained there through the forenoon. Eggs hatching? Calls being exchanged?

The hen's spread-wing position on the nest indicates she may be covering more than eggs. The weather (cold rain) would be reason enough to offer her tent to her babes. The gander remains close by. In late afternoon, during a short break from showers, my binoculars reveal a yellow body, maybe two, bobbing in the feathery nesting materials. Good to know of success.

MAY 16

The weather has cleared from gloomy wet to sunset relief. From my low deck this evening, with gin and binoculars at hand, the "best place in the world" brings Great Meadow's wonders stage front.

A fisherman paddles by in a kayak. I see he has learned well how to spend his leisure hours. With unlit cigar clenched between his teeth, he silently maneuvers his craft from one bushy hummock to the next, where "that big bass" may be waiting. His paddle moves the boat silently, his hands having been schooled to be gentle when placing the paddle back upon the boat. His casting is proper, sending the lure where aimed.

He understands the pleasurable depth of this pastime. Great Meadow accommodates his precious moments with stilled waters and a simply magnificent sunset. The "big one" is secondary.

My observation is periodically interrupted by a female northern oriole's nest-building activities. Too bad I've missed her site selection from the start, as I would have paid rapt attention to her sack nest construction.

What I have seen that differs a bit from my last words about this action is this: A pendulous nest has already been formed with walls still thin enough to see her moving about inside. This early evening's newer choice of nesting materials is dead grass stems that I've seen her plucking from the wet edges of Great Meadow. The stems vary from six inches to twice that length.

She actually descends headfirst into the bag. The entire nest jiggles as she goes about her grass weaving from the inside. Although she plunges

A flicker at its nest

headfirst to sew, I can see the tips of her tail feathers just out of the nest, and I smile at the wiggling—proof she is weaving.

Adding to these special moments are two tree frogs loosening up their vocal apparatuses before nightfall, and a scarlet tanager wheezing its call from some station. Hummingbirds are letting me know that their patrolling paths through the spaces of my observation "office" must be kept clear.

During all this, male red-wings down on the marsh exchange yodels with their contenders, while robins flute their presence in the great oak. The phoebes, alert on garden posts, allow no insect worth its nutrients to cross within their zone of eyesight; all reminders of my good fortune of being in the midst of Great Meadow's variety of life.

MAY 28

Memorial Day

My father served his country in World War II; I served four years in the United States Air Force during the Korean Conflict. Every year, I decorate my family's graves, along with those of my great- and great-great-grandparents. This is an annual offering that comes complete with gratitude.

There are other days, similar in a way, that emanate from Great Meadow to me, having to do with its history and those who have had a hand in it.

In the beginning of this journal, I shared the story about how the early homesteaders had been dependent upon the grasses growing in the Meadow to feed their livestock—really to keep them alive. On these Memorial Day callings, I also drop by the grave of one of those homesteaders. I offer no real thoughts, but I feel a deep reverence as I leave, knowing that in those brief moments, Great Meadow's history has touched me once again.

Now back at my viewing station, my eyes follow the action of eight Canadas flushing out of the thick grasses that cover the far cove. Their flight covers only one hundred yards, but lifts them out of the thick greenery over open water. At that point, all eight geese cup wings, dropping to a quick landing. Their conversations immediately cease.

Clearly, there is a purpose to their short hop. Ah, bath time! Every bird flaps and dunks together, their vigorous bathing lasting close to five minutes. With bodies refreshed, the gathering turns and swims back into their green world to continue the afternoon's dining in the sanctity of grasses.

JUNE 1

There is difference between a worry and that of holding a deep concern for the planet-threatening problems we all face. Great Meadow is a living, active, many-layered organism unto itself, a provider for its diverse multitudes. I am one part-time observer who tries to learn from what I see. As each year passes, especially during the seasons when I remember wildlife activity across the marsh to have been nonstop, periodic absences in species and persistent low numbers toll loudly in my brain.

This day, though, sparkles. The expanse of Great Meadow's mid-spring colors catches and holds an appreciative man.

A female hoodie swings into view from behind a buttonbush thicket. Adhering to her today are nine balls of fluff, each ball equipped with bobbing head and automated paddles. No matter their mother's movement, every duckling repeats it; nine times the joy to watch.

Later, a wood duck hatching of eight repeats the scene: mother trailing automated ducklings. She seems to be headed for an abandoned muskrat lodge, perhaps to give her charges a chance to rest.

I glance ahead and there, blending almost perfectly with the gray mud of the lodge, is one of the Meadow's more private and solitary residents, a huge snapping turtle. It has mounted the lodge to get a massage from a few rays of sunlight. Turtles, like other sunbathers, are apt to succumb to the sun's therapy. There it wilts, obviously injected with the sun drug, its massive head and long neck hanging limply over its perch, totally relaxed—or so it appears.

After long years of watching nature, one understands that her creatures possess a seventh sense. Now, this turtle is about to exercise Sense Number Seven. The mother wood duck and her company of eight continue a course headed for the muskrat lodge. She knows its location as a stopping place well—but today, the lodge is a snapping turtle. My eyes are unable to blink.

When the wood duck family is within two feet of the lodge, that inner sense ignites the snoozing predator. The great turtle literally catapults off the lodge into the water, its long neck aimed directly at the mother wood duck. When the body of the turtle hits the water, it creates a tsunami of sorts, launching a wave that raises the entire family of ducks above and just out of reach of the snapping jaws of the lunging turtle.

The mother duck flushes instantly as the great wave literally flings her entire clutch of innocent babes into a clump of leatherleaf, its stiff stems thwarting the turtle's advance. The turtle disappears off into the marsh.

The ducklings scramble out of the cluster of stems and realign themselves behind their waiting mother, leaving behind what might have been. For the old turtle, the missed duckling lunch was only that. When one lives for a century or more, what's one missed lunch?

JUNE 2

The pied-billed grebe's call pleases my ears before early rising time. Each year, I hope its calls are answered. They seldom have been.

JUNE 3

Aquatic waterfowl count (ever decreasing):

* Hooded merganser, one female trailing nine

* Wood duck, one female trailing eight

In previous years, at least six broods of each species have paraded young ducklings. Just over a decade ago, twice that many was the norm.

Thoughts about tree swallows return today, following a conversation with a neighbor who is concerned because only two of his many swallow houses have been chosen as nest sites. Again, the thought that things aren't right burns through my head. Only last week, hundreds of swallows were swarming over the marsh in their quest for gnats—but not a swallow is flying today. Where have they gone?

Traditionally, hundreds of migrants pass through, and in those flocks fly swallows that were hatched here at Great Meadow. Those should stay, find nest sites, and raise young. This year, for the first time, no birds are knocking on any of my seven swallow nesting boxes.

JUNE 5

Evening time

Six to eight chimney swifts over the marsh. I have great admiration for this species, having once lived in an eighteenth-century home with a chimney that housed a family of swifts. Truly, they are boarders of high interest. Their nests are literally glued to the bricks inside the flues of old chimneys. The parents use some concoction of their saliva as the adhesive, not only to secure the nest to the bricks, but to stick its building materials together.

The nest-bound swiftlets make eerie squeaks and whistles every time their parents bring food down the chimney. Those sounds reverberate throughout the entire homestead, causing guests to squirm a bit until the story is known.

JULY 3

Summer rests throughout Great Meadow. The color green, having already paraded through its spring chart of delicate shades, has now settled into its summer wardrobe. The riotous business of spring's styles has also left the stage. Now, as the usual spring wildlife hustle subsides, my eyes must search for activity.

The few birds that are nesters in the marsh's watery habitats stay mostly

Bladderwort - many suction traps seeking zooplankton

hidden, since motion reveals their incubating stations. Even the red-wings hesitate to flash their epaulets, doing so only if there is a perceived threat, such as a neighboring male that may venture too close.

JULY 6

Midday heat puts a hold on most animal activity, although great blue herons seem to be exceptions. Sunbathed, mid-marsh perch sites are favorites of that long-legged crew of fish stabbers.

Day after day, I notice a particular heron on its perch of choice. A closer look may find others equally occupied, all erect and motionless, particularly when it's the time of day to relax and allow for the morning's catch of fish to be digested.

Many times what follows is the serious business of preening their glorious feathers. Feather tending, for all birds, is a necessary chore. With great blue herons, who have endless types of feathers arranged along a three-foot span of their upper bodies, the activity of keeping them all clean and flight-worthy means barbering every day.

Take the hair-like wisps of breast feathers alone, which apparently require repeated cleaning and grooming, with the heron's neck severely cranked upward and its beak thrust straight down alongside his chest. Much attention is also given to its wrist feathers, which the heron strokes one by one with its beak in a quick upward combing motion. (Strangely, the bird seems to pay far more attention to the feathers under its wing than to the secondaries and primaries on the fingered wing itself.)

Roughly midafternoon, on slow, hot days, some exercise seems to be in order for herons. The game? Seek out another heron, challenge, and give chase. During the pursuit, emit a loud squawk or two to alert the marsh of your whereabouts and intent. Play the chase game until satisfied that the pursued has been put in its place.

With the heron action over, the marsh returns to its summer siesta, leaving observers to admire its floral gardens with their endless waterscaping. In one place, the surface will look like a grand, flat bouquet, presented in parts, with stretches of water lilies that float multi-spired, lustrous white blossoms upon sheets of green lily pads.

Not to be outperformed, the spatterdock family of lilies, with bold, cupped blossoms of bright yellow each held high above its own floating pad, shows long drifts of its unique flowers. In the more delicate gardens, where the parent plants float free, one may see the tiny yellow, orchid-like blossoms of bladderwort, or a drift of pink water-willow.

When at a marsh during garden display time, one cannot ever direct his or her attention to one thing. Dragonflies may be everywhere—brilliant, bold insects, released from their underwater nymphal prisons, now free and in control of the sky.

It may be said that the dragonfly's command of flight exceeds that of all winged creatures. Their sly, delicate cousins, the damselflies, flutter instead of fly, tentatively reaching out to perches, landing with the delicate touch of a fairy. Dragonflies carry polished, radiant colors whereas damsels, soft like their namesake, are lightly touched with pastels and maybe a dot of color, looking as shy as their carriers.

JULY 21

Looking out over the Meadow in late afternoon, the heavy, moist air carries few sounds save those of the recurrent protests of the herons.

When one thinks a bit about the bird clan, it seems that vocal cords that evolved to issue lovely notes are pretty much restricted to smaller birds—so we call them songbirds. Certainly vultures of all varieties, like the herons, are limited pretty much to grunts of protest. Parrots, birds of prey, all the larger sea birds, even penguins, produce nary a sweet note in the bunch.

SEPTEMBER 5

Hummingbird family heads south.

OCTOBER 27

All of September and most of October have passed. Meantime, even while knowing that the numbers of stories, scenes, transitions—even once-in-a-lifetime moments on Great Meadow—are slipping by during this climate change, the old in me seldom lifts my pen; a troubling stretch again.

I do give fleeting thought now and then as to what age has wrought, but thankfully distractions of the positive ilk snatch them away. Depression is unknown in my life. My situation—active, much loved by Barbara, my fine home in the "best location in the world," beneficial work through my Russell Farm and Forest Conservation Foundation, and overall good health—leave no room for the gloom egg to be laid, let alone hatched.

OCTOBER 29

My pen only asks this question: Here at Great Meadow, diversity is falling off a cliff—but those silent changes go unnoticed; who is paying attention?

OCTOBER 30

Hurricane Sandy is visiting this date, her Halloween trickery in the ferocious form of blowing winds and flooding seas raising havoc to the south. Her windy fingers apply her test to two of my largest hemlock trees, putting them into a permanent sleep directly across my driveway.

In the lady's gale, with her horizontal rain, 150 or so Canadas, even while being flipped and pushed sideways and back, manage to enact a beautifully controlled landing on Great Meadow's white-capped waters in these strong crosscurrents of wind. From where did those waterfowl come? Why are they flying in the swirls of Sandy anyway?

They are here, safe—but the birds seem jittery. They are split into groups, one group skittering a short distance across the water, followed by another and another. Their restlessness is not surprising. Unusually low atmospheric pressure accompanies hurricanes like Sandy, and such conditions unnerve wild creatures. Animal activity slows during times when barometric pressure drops, as it will during stormy weather. Adding to their fidgets is the possibility of gunners lying in wait, as this is waterfowl season again.

OCTOBER 31

Sandy has hurried away. She severely tested the area's trees, one by one. Most passed her harsh gusts, but those trunks, limbs, and roots unable to meet her demanding winds were dealt with harshly. In essence, hurricanes clean nature's house of weaklings.

A barred owl is perched on a buttonbush limb in mid-marsh this early morning. He looks oddly bedraggled, feathers totally drenched. He keeps shaking his body to rid himself of as much feather-water as possible. Perhaps he met some unfortunate encounter with the hurricane. More likely, his waterlogged condition resulted from a predatory attack upon some prey like a muskrat or a mouse that had been swimming in the marsh.

NOVEMBER 10

November could not unveil a finer day. Great Meadow's mirror reflects pure blue. A nudge from some unknown source gives me a reason to walk the forested edge of the Meadow, north from the house. The broadleaf trees are on vacation, but each conifer continues to dine on the sun's lowering rays.

I drop by my old, patient canoe by the marsh's edge. Close to— no, over fifty years, I've been accommodated by what I know as "this old friend." Oh, the journeys it has floated me on, the explorations up rivers and into hideaways created by those same beavers that build gates across their private waterways to discourage most trespassers. We've plied hidden ponds and many dark yet enticing brooks, and the fish stories we harbor together, confident in their veracity.

Like me, my canoe has become creaky, but ever ready to float my old bones onto Great Meadow in quest of one more moment to add to our long relationship. A meaningless gesture of course, but while I spend a minute here looking out across "my" Meadow, I place my hand on my old conveyance and know what it has given me.

The Meadow's edge just north of the house presents a habitat where glacial boulders are stingy about offering rootholds to trees. Deciduous trees have met those challenges, together producing the finest stand of "pure" trees I have seen. Silver birch, red and white oak, red maple, and basswood, all have stood in company, high and strong, for at least a century. Walking through those strong trees, with their understory of massive boulders, always softens the edges of any challenge I may be facing.

Ahead, at the wetland's far north end, comes soft chatter from what must be many geese.

They come into view as I step up onto a hummock close to the edge of wet, sunlight reflecting off 150-some birds. I know them; these are the battered birds of Hurricane Sandy. They are recuperating here.

Their conversations are like whispers mixed with low groans and wheezes. The occasional goose exercises powerful wings, then settles again. How fortunate to simply be standing here in witness to their company. The birds have come from somewhere to this refuge, resting here as Hurricane Sandy swept her spiral arms of wind over the land.

NOVEMBER 12

The birds continue to hold in the marsh as a second northeast storm drenches and bothers with sleet, rain, and limb-snapping winds. I'm sure they have "felt" the weather coming and held in the shelter provided by the Meadow; no wasting fat reserves by flying in taxing weather.

During soft-talk, they recuperate in today's healing sunlight.

NOVEMBER 17

Adult bald eagle: perched.

Again, there is great morning entertainment. The large "hurricane" flock of resting Canadas departs in what appears to be prearranged squadrons, split into three groups. Three or four minutes separate the staggered departures. Within five minutes, the entire mass of birds is gone. Watching how the still-persistent wind gusts disturb flight formations, and continuing to watch as individual birds alter their course to re-establish proper places in line, keeps my attention.

NOVEMBER 28

"Sandy's" geese leave every morning. They keep returning, but their final departure date is a secret.

Their space of open water is shrinking by the day. Last night, ice closed in to the point where open water space was insufficient for all the geese. The geese stayed the night, but many simply stood around on the ice, not straying far.

Morning's light comes. By watching the geese and loosely timing the action over these many days, seven-thirty to a quarter to eight is, reliably, the departure time for the first group to take wing.

One forms opinions while repeatedly watching some activity in nature. I'm beginning to think that this collection of Canada geese is composed of at least five separate "clans." The members of those clans stick loosely together while relaxing.

With the approach of morning takeoff time, there is a general shuffling of birds throughout the various groups. When the notice to take flight is acted upon in, say, Clan A, birds belonging to that clan take flight in concert. Scant minutes pass, then Clan B birds are up and away. Clans differ greatly in numbers, with one having only ten birds and others maybe forty, fifty, or more.

This staggered flyaway action, clan by clan, empties the marsh of birds every morning. There does not seem to be any regular order to their leaving—meaning some mornings, the clan with the largest number of birds may be the first to depart, and other mornings, the final group to leave may be the largest.

One more observation about this particular collection of migrants: Without exception, after the last clan leaves, two geese, perhaps a pair, will remain on the ice. Following the great clamor of beating wings and verbal exhortations over a quarter hour of mass evacuation from Great Meadow each morning, those two geese, necks fully extended, graceful heads alert to their world, stay behind. After roughly a minute, they also leave.

NOVEMBER 29

A cold night. All reaches of open water are now closed, and the marsh wears its winter cap. I think the geese knew of the freeze in advance.

DECEMBER 8

Once again, I learn: Never resort to assumptions when nature is the subject. Two days of 50°F air temperatures, and Great Meadow's roof of thin ice has lost its grip.

Late yesterday afternoon, just at dusk, a familiar sound came to my ears through the day's thick fog. Believe it—the flock of Canada geese was returning! Their wonders never cease. Surely these are the same birds that have occupied this wetland for the last weeks. What special sense told them that open water had returned to Great Meadow and would welcome them? Where did they fly to ten days ago?

Cardinal (male) at feeder! A first here.

DECEMBER 12

I feel very troubled where the local climate is concerned. The weather is more spring-like than December-like. The Meadow skims over with ice one night, only to let go of winter's touch the next day. This morning, I find that Great Meadow has only about half the number of geese I'd seen late yesterday. I didn't hear them leave.

The northern cardinal is seen, but he comes only during the morning hours. His flashy apparel gives him a certain aura. The red-bellied woodpecker is a regular again. The family of chickadees outnumbers those over the last two seasons. The goldfinch flock seems to grow daily, with forty to fifty here regularly. Hairy, downy, and pileated woodpeckers are regulars, although the pileated sticks to his grub-searching in the oaks, never braving the suet offering. Titmice, white-breasted nuthatches, and blue jays keep my eyes busy. The ravens swoop in most mornings, grab what is offered, and flap away to play games.

DECEMBER 13

The saga of this year's comings and goings of the Canada geese has rhymes and reasons known only to them. Near midnight, I hear a flock of geese return. This time, their voices are barely audible. That is different, as usually flocks leave no question as to their whereabouts after landing, especially after hunting season has passed.

DECEMBER 14

At dawn, it's clear that temperatures have fallen enough for Great Meadow's surface to freeze again. There, out in center meadow, are the geese. Most are crowded into a small reservoir of open water, with the balance standing close by on the ice. (If the Meadow's surface was frozen over when they returned last night, could the iced-over condition have changed their voicing pattern?) Come breakfast time, all the geese are up and away.

Now, what's that? There, not far from the shore, a single goose is afloat in barely enough open water to hold her. Her head is tucked under one wing and her back is encrusted with a heavy layer of frost. Is she dead?

At nine o'clock, the bird's head is erect. She is moving in an agitated way. When she wrenches her body, I can see that overnight, part of her left wing froze into the new ice. Struggle though she might, the ice holds fast to the wing's primary feathers. Clearly, the goose has been injured. Obviously her situation is dire, and hard to watch.

By midafternoon, she manages to free her wing—but the air remains frosty enough for the surface of the water around the goose to keep freezing. To counteract that threatening situation, the goose keeps thrashing her wings and turning her body to stir the water. Darkness closes the scene.

DECEMBER 15

At first light, it's obvious the struggle has ended for the goose, but not without one final effort to survive. Somehow, during the night, the goose managed to pull herself out of the water and onto a patch of mud. Light reveals only a frosted lump.

DECEMBER 17

The flock of geese has returned the past two nights. Those who find room in the open water squeeze there together, while the rest stand around on the ice as before. They've stayed each day, continually conversing in barely audible voices. They move little. What information is being exchanged? Certainly vocal chords do not evolve to make only noise.

Unlike yesterday, but following a pattern used on previous mornings, the entire mass of birds takes flight, again in staggered groups. This time, no birds remain on the ice.

As the last clan takes wing, it circles the Meadow's perimeter. While I watch, a single goose drops away from the formation and returns to the ice. Perhaps some injury has made him unable to keep pace.

Now, at midday, that single goose remains. Could this be the mate of the lone goose that died on the Meadow? Did this goose notice that his mate was not in the final skein where she should have been, and that is what brought him back onto the ice?

Overnight, the dead goose on the mudflat has been ravaged. Fox? Coyote?

DECEMBER 18

An early morning look reveals the single goose still on the ice. He moved little during the night.

Finally, with hesitant steps and intermittent pauses, he makes his way towards the patch of open water, where all the geese were yesterday. When he comes to the edge of the ice, he stops, standing perfectly still, head held high, calling, calling, calling.

I watch as he steps off the ice into the water. He turns into the slight breeze, hesitates, and then flies off in the direction his flock took yesterday.

Those few feathers on the mud; why do they carry me back to my years here, and my curiosity about Canada geese and, yes, to the pair that had chosen to build their nest so close to my watching place? After raising their hatchlings, that family joined the others here at Great Meadow and I lost them to the crowd. That separation does not lessen my hopes for the future of "my" geese.

When an incident happens, like the pair that stayed on the ice after their flock left, or when that single goose separated from his skein and returned to the ice—calling, calling, calling after I had just watched the injured goose die—I know I must not make a connection.

That—the not-knowing—keeps my attachment at the sentimental level, where it always must stay.

DECEMBER 20

An observation: The space of open water recently used by all those geese does not seem to freeze over, despite several nights of well-below-freezing temps. Perhaps it is reasonable to think that the long and active occupation of two-hundred-plus geese has created conditions that could explain the open water.

The space, roughly two hundred yards long and thirty yards wide, held geese that constantly fed on aquatic succulents. Those many beaks pulling at underwater vegetation would have caused a major disturbance in the substrate. That disruption would then create new layers of food for fishes and crustaceans, and especially organics upon which macroinvertebrates—the gnawers, scrapers, nibblers—could forage. All that eating would create copious amounts of detritus. Add a deluge of goose scat from two hundred intestinal tracts to that underwater soup, and the resulting decomposition would create a lot of heat.

Today, a pair of black ducks is feeding in that open water space. (They shouldn't be here this time of year.)

DECEMBER 28

Snow: seven inches.

DECEMBER 30

Snowfall again: four inches.

DECEMBER 31

Weather in the twenties with snow cover.

Good to have a sense of winter; I enjoy the bite of a northwest wind. This morning's ambience is giving my memories a chance to drop by and be appreciated again; another year.

Beaver Lodge, insulated for the long, cold haul

2013

Ravens kraak. Muskrats whine.

JANUARY 5

A male coyote. It shows up during mid to late morning, walking slowly on the Meadow ice, meandering, which is not its regular hunting technique. Usually coyotes keep focused straight ahead until their noses turn them towards some scent. This animal moves from one emergent, aquatic bush-clump in the ice to the next; checking for mice, I expect. It goes to this year's muskrat lodge and digs a bit, but apparently its nose indicates that no further effort is worthwhile.

JANUARY 6

Today, the coyote (no doubt the same animal I saw yesterday) is out along the far shore. It's a sunny day so, as any animal might under the effect of a warming sun on a chilly morning, it pauses and enjoys the moment. With no threats or hints of food arriving on the wind, and with no urgent pack obligations, rest time is a rare treat. A sit-down is in order and an itchy ear must be attended to. The itch not easily defeated and the scratch is long. Tiring work that, so why not succumb completely to the warmth offered by the sun and have a good nap?

I've seen many coyotes, but none so relaxed; a high-value moment in which to see a wild animal in complete repose. It rests for about two minutes, and then returns to a sitting position. The woods beckon. The coyote complies.

Odds for a long life are against coyotes that use Great Meadow's territory as a home base. It is good this coyote can't know what I know about its human enemies. A farmer's son I am acquainted with brags about his hunting skill and the numbers of coyotes he has shot. And then there are the trappers.

JANUARY 9

Bluebirds at Great Meadow! On the south edge of the Meadow's cattail colony, I count at least ten of these delightful blue singers actively feeding on some type of food that seems to be available at the base of the cattail stalks. The birds balance for a few seconds atop the seed heads, and then drop down to grab some unknown food.

Bluebirds in January? The Age of Warming brings dark pleasures.

JANUARY 13

Without traditional cold, it has been easy for the beavers living in their restored lodge by the Meadow's brook outlet to maintain open water around their living quarters. Today, midafternoon, three beavers are out on the ice, each munching on water lily rootstocks they've retrieved from the pond's mud. Unknown to the feeding beavers (I expect) is a mature bald eagle, perched on a limb only twenty feet above the picnickers. Its gaze zeroes in on the unsuspecting "food," a mere dive and grab away.

Once an eagle reaches maturity, obviously many of life's lessons have been learned. Even if the eagle's stomach is empty, I expect it knows that something very bad would happen if it were to sink its talons into this prey. In the event of an attack, the beaver would drop off the ice into the water, carrying the eagle along. No good would come of that. Yet the possibility of a meal a mere talon-grab away, or some other unknown reason, holds the eagle's interest all the time the beavers indulge in their winter's lunch.

JANUARY 16

60°F!! What thin snow there is struggles to keep its traditional white.

JANUARY 17

Four inches of welcome snow this morning, with more eagle shenanigans around noontime. This performer is an immature bird, standing on the ice shelf that rims the open water around the beaver lodge. My binoculars show a beaver-peeled stick floating in the water. That stick seems to be of interest to the eagle. Why? As I watch, the bird *hops* into the water, great wings splashing as talons reach out to seize the stick!

Now back on the ice, the eagle appears to try and get nourishment from the prize. It pecks at the stick again and again. The piece of wood is about fifteen inches long and one inch in diameter, and has been completely stripped of any bark by beavers—yet the eagle keeps turning the stick end-for-end with its talons, and working it with its beak.

With nothing gained by that exercise, I am surprised to see the eagle repeat the same action, twice, with different floating sticks. Could it be this bird considers those light-colored objects a possible food source? That gives little credit to the eagle, but what alternative explanation could there be, other than it's just looking for something to do?

Moments later, the eagle flies to the top of the beaver's lodge, grabs yet another peeled stick, returns to the ice, and repeats the stick-turning yet again. An hour later, a second immature eagle coasts in and lands on the ice one hundred or so yards from the "stick eater," who then joins the new arrival.

FEBRUARY 11

A male Carolina wren, first seen in 2001, has returned. Here's yet another southern bird reacting to a warming climate. Wrens are feathered bundles of unbridled energy, each species complete with repertoires of bubbling songs. They disguise their pugnaciousness with a spirited verve for life.

MARCH 6

Hurrah! After many years, winter has lived up to its traditional New Hampshire definition—that is, several months of cold and snow. It's early March and nearly sixteen inches still blanket the land. Snowshoeing has been limited due to the texture of the snow, along with some complaints from old muscles.

Bird feeding is costly this year, a sign that there are many hungry visitors who find eating at tables set with seeds an easier option than scrounging for food that is always hiding. The local ravens provide excitement every morning as they play tricks with their food.

When I go outside to feed the birds, I always wear the same coat and hat, and I always repeat the same call words: "Come on. Come on." Although I never see them, I know the ravens are watching.

With the ritual complete, I return inside. Within five minutes (the time sometimes varies), a black shadow will catch my eye as a raven sweeps by my windows. How quickly they come to the food depends upon the competition for the standard fare of peanut butter on bread from others.

The ravens aren't the only birds who have learned about the morning handouts. Chickadees, tufted titmice, and nuthatches are waiting. They fly to the bread almost immediately and begin to feed. They don't present much of a threat, but the squirrels, both red and gray, often race to the slices, grab one, and abscond with the food.

Ravens are ever cautious. They never land close to some perceived food. Once on the ground, they have many "look-arounds," and their strolls always take on an air of absolute indifference. The final moves on their "be careful" list are sidling up to the food while leaning away and then, if nothing seems strange, making a flash-stab at the object. Suspicion still reigns, as the ravens never grab the food on the first two or three stabs. When they deem the opportunity safe, they quickly snatch the parcel, their wings already spread in escaping flight. I expect this cautious behavior has evolved from past encounters with bait set in traps or snares meant for furbearing animals.

MARCH 9

Twice this week, I've observed a pair of ravens share food at a location well away from where they took it. The male always snatches the food and shares, but there is a process to his sharing. Both of the incidents involved slices of bread covered with peanut butter.

Yesterday, the male grabbed a slice, set his wings, and glided to a landing on the snow-covered marsh a short distance from where his mate already stood. Upon landing, the male held the bread with his left foot and ate about half the food. He then lifted his head. With that apparent signal, the female hopped across the snow and faced her mate. The male then flew off and female took her time finishing the food.

This morning, I watch a repeat performance. The male takes the food and lands on a large, dead branch of the Old White Oak. Again, he flew to where his mate had been waiting. The male eats. She watches. When the food is about half gone, she hops to the male along the branch—but now, because the limb is not wide enough for her to pass, she simply hops over him! Then, as before, she turns and faces the male, and he leaves the rest of the food for her.

Two incidents of sharing; dare I write the word "magnanimity?"

MARCH 13

Red-wings returned this morning, along with three common grackles. This afternoon a turkey vulture arrives.

MARCH 18

White-throated sparrow and twelve inches of snow (in the woods, there is generally three feet).

Again today the raven pair cannot be ignored. The male has made several trips to my offering of roast turkey leavings, which I'd tossed away onto the snow. An hour later, I see him and his mate out on the marsh snow, and the pleasure of watching that pairing begins again.

Today, they are doing what can only be described as the "raven hop," a series of hops and skips that the pair performs more or less together. Every fifty feet or so, the male stops, convulses his body, and regurgitates what must be turkey. He then offers the result of this reverse-swallow as a snack for his mate! In an action that can only be described as ever so gentle, she takes small bits at a time from his beak until every morsel is gone. The male shows complete patience with her nibbling.

Now, with that depiction of compatibility completed, the pair hops along another several yards and stops. More turkey is offered and accepted. Twice again, this raven-style sharing scene is repeated—but just before the last stop, the male realizes that he'd apparently dropped a bit during one of his hops. His lady notices, and hops back to snap up the crumbs.

What do these actions show? To me, each one of these ravens clearly displays a spirit of congeniality, behaviors that suggest far more than some natural linkage existing between them.

MARCH 25

Canada geese fly by, a pair. Within a minute, they return and land on the marsh snow. I always hope the first pair to arrive is the one that commonly nests here.

APRIL 1

A mature bald eagle perches on a low stump, and in his company are three great blue herons, all within twenty feet of one another. Unusual proximity for those birds, but I expect a slice of open water presents an opportunity for food.

At the upper end of Great Meadow, a wide patch of water has opened and attracted a couple dozen geese. Their conversations carry into my bedroom and bring relaxation to my night. Adding to the spring's pleasures, last night coyote music touched this dozer's ears.

APRIL 3

40°F.
Five tree swallows.

APRIL 7

Just as the sun is setting, one small opening in the ice offers a rather dramatic pageant: Two hooded mergansers and two male common mergansers are literally swimming under the stare of a great blue heron. I believe I'm watching forbearance on display by the tall bird with the dagger beak. But then, herons don't relish duck, and the ducks know it.

APRIL 11

Three ospreys are spying for fish from above the open water spaces.

APRIL 15

A phoebe returns with a bump against my office window, snatching and ending the day for a spider. I feel close to this bird. My small garden is just outside my office door and when I work there, the phoebe is often my companion and supervisor, sometimes perching on my hoe handle when I take a short break.

APRIL 16

I'm waiting for more of my "star" species of duck, common mergansers. The earlier arrivals are fishing separately.

For the past week, the resident Canada geese have been active verbally and physically as they sort out the group's relationships. This is obviously a complex activity, and all seem to be involved. Calm reigns at times, usually when all the geese are feeding together, or swimming from one point to another. At other times, some disagreement or objection breaks out and one goose suddenly takes wing and attacks a fellow flock mate. The one

Family of Canada geese on Great Meadow

attacked typically flees, but only to a point a few yards away. That only induces the besieging goose to make its point again, with a bit more gusto, until the subject of pursuit gives up and retreats to some far point on the marsh. Satisfied, the attacking goose returns to its congregation, skids to a splashing stop in their midst, and with a flourish, stands high in the water in order to flap its wings in victory. Strange though, not a single bird in the flock seems to pay the boaster one iota of congratulatory attention.

APRIL 17

It appears that the pair of geese we have watched nest and rear offspring over the years remain pleased with one another's company. Even though they make it a point to reestablish their nesting territory, there are free-loaders without mates who occasionally cause trouble, thus necessitating repeated chase-a-ways by the male. I've yet to see him successfully challenged.

This year, we have a concern, since last year's nest deteriorated to the point of appearing unusable. I've seen the pair standing on little but mud at the old location. Will they rebuild or move to a new site? Decisions are being made.

APRIL 29

A scene with muskrats: A female in estrus and a male in pursuit are the two actors. Observing is easy, as the encounter takes place around a clump of aquatic bushes that measure about five feet across in all directions, all surrounded by clear water.

Immediately upon meeting, the muskrats begin a counterclockwise chase. A pattern quickly becomes evident. Practically every time the female swims to the nine o'clock position, she stops. The male, always close behind, takes this opportunity to quickly mount her and have what can be described as staccato sex (which never lasts for more than two seconds). Then, the female slips out from under him and continues her counterclockwise, around-the-bush posturing.

The gallant pursuit goes on for twenty-three minutes. Twice, the female briefly changes course by a few feet, at around the twelve o'clock position. The move seems to confuse her pursuer. He stops, moves a bit right, moves a bit left, and waits. When his ladyship returns to her regular route, the male gets back to the routine of copulating at the nine o'clock mark. And around they go again.

To up the ante, or perhaps to test the resolve of her mate, three times upon reaching the ten o'clock spot the female takes a shortcut through the thicket of buttonbush stems. Being slender, she can negotiate her way through the stems with ease. The male, larger in girth, has to struggle, causing him to fall behind. He needn't worry, because all three times, the female waits for her suitor! Once he catches up, they are off again, swimming backwards in time.

APRIL 31

Two cormorants. Whenever this species appears at Great Meadow, I wonder about the route they've followed to arrive here from their natural seaside habitat. What stimulus moved them? Will they reverse the same route when they return back to the sea?

MAY 2

A cold front event: Standing on the deck, I hear wind in the far cove—but no trees are moving, and the marsh is perfectly calm. After several seconds go by, I see spray begin to funnel upward in a counterclockwise spin. I realize I am witnessing a waterspout at Great Meadow, its mists revealing a vortex caused by a leak of heavier, cold air from above. The spout lifts its mist of water close to forty feet skyward, moving across the Meadow a couple hundred feet before collapsing into itself. Following this dramatic arrival, the air above the surface of the Meadow begins to cool and condense into patches of fog.

Only one cormorant today.

MAY 7

After a stint in the garden, I settle into my outside office to relax with a book and an iced coffee.

The relaxing part is short-lived. I'm distracted by a splashing sound. Water is splashed by many marsh animals, and sometimes the splashers can be identified before they are seen. I recognize these sounds because I've experienced their uniqueness three times before: mating snapping turtles. The partners are close by, and I capture their activity with my binoculars.

I've described similar scenarios before. This one-purpose rendezvous, more a tumble-and-bash-about really, reinforces my past observations that turtle copulation is performed plastron-to-plastron. The difference in the action this time is that the positioning of the female goes much more smoothly. During my previous observations, the male's efforts to flip the female into position seemed like a royal struggle. This pair seems far more skilled, or perhaps the female is more mature and ready. Here, in less than two minutes of grinding shells and raking claws, they become motionless in the water. I am amazed again at the length of time they remain completely submerged. As best as I can estimate, at least three minutes tick by before streams of bubbles are released as the turtles exhale and return to the surface.

As those wondrous reptiles go their separate ways, carrying 230-odd million years of reptilian past in their genes, I wonder: Now that the planet is turning up its thermostat, how will their order of life meet the challenges? They have certainly managed to beat the odds up to now, even during a time when the Earth was far warmer than it is today.

MAY 12

Once again I've settled into my marsh-overlook chair, binoculars and late-day martini on the railing. I sit and look over the expanse of the marsh as a blanket of low, thick fog obscures it entire surface. Such a low cloud I've never seen here before. (Usually it's water vapor that swirls and twists over the Meadow.)

What was that? I thought I glimpsed a bird. There, now, many of them—tree swallows, canting up through the fog like so many fireflies rising above a field where grasses are thick. My first reaction is gladness, because I had expected them at least a month ago—but then I quickly become distracted by the odd way they are flying. Instead of their usual circling, turning, above-water quests for insects, these birds swoop down, up, down, up, above the layer of fog. My binoculars confirm the reason to be the mating flight of mayflies!

The brief-life flight pattern of those grand, graceful insects is almost rhythmic, with its up-and-down swings through the air. The tree swallows have altered their own mode of catching supper, so in turn they follow and swallow their prey up, down, up, down.

MAY 15

Only a dozen tree swallows remain, now skimming over the water. I watch them dine just after sunrise, but then none appear over the Meadow during their usual afternoon feeding time. Have they gone already? Surely, more must be on their way, but I'm beginning to question my own expectations. Where does that leave the virtue of hoping?

JULY 20

Already, I can feel the sunlit hours losing their battle with the late July calendar. I don't like it that time is speeding up. In my youth, I wanted the days to go by faster. There were a lot of events and important dates I couldn't wait for, and it seemed to take forever to get to them. Now, Thursdays seem to follow Mondays, leaving me to wonder how I lost the missing days in between.

It is worse than that. Take gardening: It seems that I just get around to putting my vegetable canning pots, pans, grinders, and jars away, when I notice the tomatoes are ripe again, my peaches ready to pick, and the blackberries growing across Great Meadow are calling.

AUGUST 7

This morning, as I have done since coming to Great Meadow in 2000, my breakfast watching is routine—but seldom is breakfast completed without some moment worth letting the coffee get cold.

Today, the marsh's August browns are beginning to tint the gardens of aquatic grasses. The tips of the long, slender bulrushes, and the graceful, now-exploding seed heads of the barberpole sedges are wrinkling a bit, barely touched by old age. Neighborhoods of bluejoint grass sport jaunty, airy heads of tan as their autumn finery, proving that a little aging does not erase beauty.

As I watch the effects of time passing in the Meadow's gardens, my thoughts switch to last night. As I lay in bed, doors and windows open to the marsh, I listened as closely as my ears let me. The weather was warm; I was distracted now and then by the ticking of my alarm clock as its second hand answered to electric impulses from its battery heart. I concentrated and reached out with my listening to inquire. Not a single amphibian voice spoke. Somewhere out in the Meadow, a great blue heron broke the wondering silence with a lonesome "kraak." Sleep eventually came to relieve my growing inquietude.

SEPTEMBER 8

This morning, as my eyes pass over those lovely reaches of the Meadow's late summer colors, I realize another month has been announced by the plant artists who use only brown paint. Still, the Meadow retains greens and soft yellows, all holding remnants of warmer days—but I'm missing something that one always expected of Great Meadow: the voices of birds. My listening comes up empty, as does my looking. No calls, no birds anywhere across the expanse of bushes and water.

Of course, I do not expect to hear any of the usual vocalizing of the Meadow's neotropical nesters, as they, with the exception of blue jays, resort to more soft twittering than purposeful singing. But even the geese have gone mute, and the usual, occasional outbursts of quacks from either mallards or black ducks are missing.

Dark thoughts grow; I've not seen a grebe this year. No call from the enigmatic American bittern; the air around any large marsh is empty without that bird's calling. No northern harrier has cast its bouncy shadows onto any blade of meadow grass or reach of water. No river otters.

A bald eagle came in January. Strange; but maybe not, because another came on April Fool's Day, only to leave before an hour could be counted down. No fish has felt the shock of osprey talons since April. Last week, two great egrets crossed my view at breakfast; neither was seen again. And where are the air-dancing cedar waxwings?

OCTOBER 9

Again, as I try to every chance I get, I go down to my watching deck late in the day. I have my journal and pencil. It is four o'clock in the afternoon, a bit early for a martini. I gaze out across the expanse of wetland. I open my journal and frown as I notice I have made no entries for a few weeks.

A beautiful afternoon. Just after five o'clock, Barbara calls from the deck above my watching place to say that cocktails are in order. I pick up my paraphernalia, go upstairs, mix the drinks, get a few snacks ready, and join her on the upper deck. We watch over the Meadow together, but with no wings working in its sky, with no sounds moving across its water, our conversations turn to current events. Not my choice.

OCTOBER 21

The afternoon air rubs its October elixir over everything, including me. The sunlight of fall follows specific rules: Be well defined, clearly separate shadows from lighted objects, and exaggerate colors. The crisp, moisture-free air opens distinct pathways for this zestful, late-season light.

Since ice left Great Meadow in late March, the year's days and nights have been absorbed by the water body. How rich the colors of the marsh can be as its days, weeks, and seasons march along. Spring's high water brings sediments that absorb color, turning the Meadow's face the shade of dry leaves. As summer settles in, the water seems to employ the cosmetic talents of the wind to change the complexion of its face, polishing it one day so that it glistens, at other times gently massaging the surface but only to the point of brushing and sculpting slight riffles in long curves that wrap around bushy peninsulas and clusters of plants.

The season of cold must be dealt with too, when the surface of Great Meadow is grabbed by winds that twist it into crowns of white. At times, it appears that the wind gets restless with its duties.

The peaceful days of floating lily pads are over. The same is true for the water bugs that just spent the summer surfing and devouring tasty swimmers possessing less speed and maneuverability than their own. Fall's cold water will soon reach inside every amphibian and reptile, and flip their internal engines to sleep mode. Great Meadow's fishes are given a pass.

NOVEMBER 10

The Meadow is colored by November's chilly shades this morning. As the air cools, it touches the water's surface, first in the quiet recesses along the shoreline, and slows its molecules enough so they cling together, becoming skims of ice.

NOVEMBER 16

Anticipation stemming from positive memories always brings good feelings, but when those expectations fail, an empty space in the stomach area wakes up and is felt.

Pied-billed grebes have given me special viewings in November. Herring gulls have often taken time off from their seaside haunts to drop by Great Meadow and give its air a twirl or two.

But where is the rest of the wildlife? My journal entries of ring-necked ducks await an update. Perhaps missed most of all are the Canada geese with their demanding presence and constant interplay that have always filled this watcher's head with questions and wonder.

DECEMBER 1

Off to four western National Parks: Canyonlands, Zion, Arches, and Bryce Canyon. . . .

Winter ice across Great Meadow

2014

A man sighs.

JANUARY 8

I'm welcomed back to the glorious white and chill of winter.

FEBRUARY 28

I've stamped many snowshoe imprints into the snow these winter months, snow that kept refurbishing the walking surfaces with storm after storm. Great Meadow's invitations to explore had to be accepted.

Many friends joined me as together we walked its borders and learned how its plants prepared seed heads and dormant buds, where mink had been, and where a single otter had plowed and sledded through the soft snow.

Even temperatures remembered where their pointers should be on winter thermometers, staying there for weeks.

MARCH 12

In 2013, I shared that Great Meadow's south-end beavers had abandoned their old family lodge and built a new one. I approached this new lodge today with one hope in mind. Having had this experience numerous times before, I know what I hope for requires patience, patience to be tested by a temperature of exactly 0°F.

I take off my noisy snowshoes many yards from the lodge, as sound travels easily through thick ice. Very slowly, I approach the lodge. I stop at its edge. Five minutes, ten, then two or three more—there it is! A low, grunting/cooing sound of contentment: a nursing beaver pup deep in its home. Those audible expressions instantly dismiss my shivers. What's that? My ears have grown old along with the rest of me, but low sounds still register well, and I hear an overlapping of what I hope is tasty satisfaction for two: twin nursing beaver pups suckling a mother's gift of life!

I get feelings of warmth and wonder every time I hear those sounds. With them comes a deep appreciation not only for the fortitude on display by a family of animals living for five months in a cramped, dark lodge, but also for what an important role they play in the habitat this species provides for animals and plants.

Nursing time is over after about three minutes. How lucky I am in my timing. Hopefully, the parents have only another few weeks before winter's ice frees them.

MARCH 21

A single red-wing blackbird arrives as spring marks its entrance in the calendar.

MARCH 22

Two common grackles come to my feeders and spend time cracking sunflower seeds, the much-needed nourishment beginning to replenish exhausted muscles.

My snow walks continue every day I'm free, and I seldom miss meeting an elder tree, where I offer my thanks for its beauty, longevity, and service. I end by mumbling my appreciation for the long, proper winter.

2014's winter paid spring's arrival no attention. The snow has remained deep, and I continue to stamp it with snowshoe patterns. The sun, however, is always the arbiter at this time of year. Its heat has finally opened stretches of the marsh around some of the south-facing shorelines.

MARCH 25

Two Canada geese arrive while I am on the deck feeding birds. As usual, to have geese return after a long winter is thrilling. This pair is very vocal as they pass. Surely I know them. Surely they recognize me. That would be nice to believe, but not knowing is best and safer for the birds.

MARCH 30

With the ice mostly melted around the shoreline and over places that are shallow, a pair of bald eagles drops by to inspect the softening surface for winter's fish corpses.

There is fish mortality throughout the long winter. Most fish that die are consumed by the many species of underwater animals; some, however, get caught in the rapid freezing. It is those bodies that appear as the ice melts, and those carcasses that the eagles feast on.

Bald eagles have often frequented Great Meadow over my years here. They have provided many exciting, eye-worthy moments. To have what appears to be a pair of mature eagles appear pushes my thoughts to places they shouldn't be. The thoughts? How incredible it would be for a pair of bald eagles to nest in the vicinity of Great Meadow! The chance? Probably zero to none.

APRIL 6

Finally, the beavers are released from the winter ice.

APRIL 24

Not only is the pair of eagles here every day, but there have been several days when there have been four eagles: two adults and two immature birds. When they are visible, I do nothing but watch. They provide top entertainment, as well as unique interactions with nature.

This morning, for example, a very large immature eagle perches on one end of a floating log. The adult watches from a stump close by. The immature bird flies off its perch and makes two tight circles not far away and low over the water. It loses altitude quickly, splashes into the water, and grabs its target prey. However, wing strength is insufficient to lift the fish. The bird does what eagles (and ospreys) do when their talons have squeezed onto a heavy load: It uses its wings as oars.

The adult watches from the stump. It takes less than ten minutes for the younger eagle to drag its anchor of a fish to a log about one hundred yards away. Once there, it gives a mighty heave and drags its meal out of the water, onto the log. The bird wastes no time getting down to business, gouging, twisting, yanking chunks of fish one after the other into its beak and down its gullet. The feasting continues for roughly ten minutes.

The adult bird then exercises its status. It spreads its seven-foot wings, flaps three times, and drives the young eagle off the meal—but not far; the younger one simply takes the place of its parent.

Interesting: The adult bird does not make any attempt to eat; it simply stands directly over the fish. I expect there is some statement of hierarchy involved here.

After a short three minutes the adult returns to the stump, displacing the youngster who goes back to the fish. Eventually, the adult flies off, and the immature eagle spends an entire afternoon gorging on the prey.

MAY 1

I've just returned to my office from my lookout deck. For the first time this April, it's warm enough to spend a couple of hours sitting in the spot where I have been an audience of one to Great Meadow's theater since the year 2000. I don't bother to bring my journal, as I just want to sit and give time for my hobbling brain to rummage through its storage attic. I remain patient as it searches for a few dusty memories considered worthy of review.

I don't expect a surprise. I expect that my brain's catalogue of old times will trundle out visions of critical events, or special circumstances gifted to me by Great Meadow. It doesn't. When I relax a bit and my mind is able to put away its expectations, the images I then receive are exactly what they should have been in the first place: A single frame of reference—the entirety of Great Meadow. That image snaps me out of my retrospective mood and brings my concentration back to exactly where it should be, to Great Meadow itself.

In this year 2014, Great Meadow does not need fixing. It is a natural complex of many living parts, a restaurant for one. It is fully stocked for every species of diner that should be here this spring, no reservations required. Since ice-out time, I have been waiting for the Meadow's annual residents, transients, and even accidental visitors to create the usual spring whoopee and party-making on, in, and over its surfaces.

I sit. I so want to greet the Meadow's guests with my eyes. I watch. Often in my journal, I've mentioned my love of the common mergansers. It is well past the time for them to grace the Meadow. Hooded mergansers? So many times I've been enthralled by their on-and-off nesting interactions. Has either species been seen? No.

Yet across the Meadow, spring is smiling and decorating as if nothing is askew. Her tree blossoms of red reflect their usual promise of rebirth throughout the local plant and animal communities.

Just now, paddling one way, now another, two adolescent wood ducks ask for attention. They get it, but it appears they are having difficulties figuring out what to do with their handsomeness.

Where are the mergansers? Why don't they come?

MAY 2

Midmorning, I return to my special sitting place. Little there to catch my eye, so my thoughts darken a bit.

I reflect that Great Meadow and I are partners in life. I give it attention and thought every day. In this journal of moments, I've chronicled what I call an "otherness" that is hindering the Meadow from being its prime self.

Though Great Meadow is but a speck on Earth, it may be unwise to dismiss its diminutiveness and not think of it as a harbinger of a profound crisis on our planet, the magnitude of which may be beyond comprehending, let alone rescuing.

Great Meadow is a living entity. It had a birth. Its form keeps changing. As has been noted in this journal, Great Meadow, along with its vast and varied dependents, is experiencing degrees of incompletion. Its deficiencies are directly related to an atmosphere that is growing increasingly warmer.

There is an invisible, mostly man-produced blight in the air Great Meadow breathes, one that I might compare to a primary feather going missing on one wing of a migrating Canada goose, or maybe a hummingbird. Hardly noticeable to the bird, but the efficiency of flight lost due to that single, missing feather will be a harmful drain before that bird's long journey is over. With Great Meadow, the journey is longer, the exterior afflictions far more subtle, and little by little, the multitudes that depend on its biological richness will diminish, are diminishing, have diminished. Great Meadow is a victim.

Still, Great Meadow remains a life-giving body unto itself, biologically rich beyond imagination—but more species of fauna seem to be moving elsewhere than are being replaced with newcomers. Shifting weather and temperature changes are forcing adjustments, both internally and throughout its vast connections, especially with birds. As a dedicated watcher, I'm finding too many days without them.

Nowadays, when I sit in my cherished under-the-deck office, where Great Meadow lies innocent before me, typically my sole diversion is a martini.

There is seldom cause to bring my binoculars to my eyes. When Barbara joins me on our upper deck for our five o'clock repast, the usual diversions created by Great Meadow's regular occupants are missing.

MAY 20

This morning's misty, soft scene with its embodiment of life brings joy and enchantment. I watch a few late-migrating tree swallows—how wonderful they are. Canada geese, a dozen maybe, are talking over one another's comments. I sit as an eavesdropper.

After breakfast, I shall make my way down to the edge of Great Meadow and give the deepest thanks I am capable of giving. While my journal now holds empty pages, the Meadow still presents the wonder of itself, making my life a bit better every time I glance its way.

White water lily with pollinating black flies

ACKNOWLEDGMENTS

IT WAS THE EVENTS THAT by sheer chance came before my eyes at Great Meadow that prompted the entries in this journal. The interpretations I've made are based upon a long life of contemplating answers to constant questions, questions prompted by my persistent curiosity.

During my years of formal teaching, and many more of mentoring, it was eager bright-minded high school and college students who helped to shape many of the thoughts expressed in this journal. Each one gave an extra dimension to my days.

Any thoughts of publishing these pages came only after I had drafted my scribblings and, with considerable hesitancy, asked if the following people would consider scanning my unedited pile of papers: Lisa Brooks, PhD, Associate Professor of English and American Studies, Amherst College; David Carroll, eminent artist and author, School of the Museum of Fine Arts Boston; David Anderson, writer/naturalist, Society for the Protection of New Hampshire Forests; Cindy Stave, designer; and Joel Stave, software engineer. Not only did they take time to read my words but, to the person, strongly recommended that I try to publish.

It was the professional editorial skills of Laurie Peyser of Atom Pie Publishing who also recognized a purpose for this journal, and gave flow and voice to the manuscript.

Gordon the watcher